Texas Waterfowl

Number Forty-six: W. L. Moody Jr. Natural History Series

Texas Waterfowl

William P. Johnson and Mark W. Lockwood

Texas A&M University Press
College Station

Manufactured in China by Everbest Printing Co.,
through FCI Print Group

First edition

This paper meets the requirements of ANSI/NISO Z39.48-1992 (Permanence of Paper).
Binding materials have been chosen for durability.

Library of Congress Cataloging-in-Publication Data

Johnson, William P., 1969–
Texas waterfowl / William P. Johnson and Mark W. Lockwood. — 1st ed.
p. cm. — (W. L. Moody Jr. natural history series ; no. 46)
Includes bibliographical references and index.
ISBN 978-1-60344-807-9 (flex : alk. paper) —
ISBN 1-60344-807-1 (flex : alk. paper) —
ISBN 978-1-60344-820-8 (e-book) —
ISBN 1-60344-820-9 (e-book)
1. Waterfowl—Texas. 2. Waterfowl—Texas—Identification.
I. Lockwood, Mark. II. Title. III. Series: W. L. Moody Jr.,
natural history series ; no. 46.
QL696.A52J68 2013
598.4'1—dc23

2012017276

Frontispiece: Male Wood Duck, photograph by Raymond S. Matlack

For Bridget and River, thanks for your patience (WPJ)

In memory of Larry Semo and Terry Savaloja (MWL)

Contents

Acknowledgments

We are very grateful to the many photographers who have graciously allowed us to use their photographs in this book. Those individuals are Trey Barron, Steve Bentsen, Tim Cooper, Greg Lasley, Ron Lockwood, Ray Matlack, Frank Rohwer, and Larry Semo. Their wonderful photographs are the result of extreme dedication and countless hours in the field. We are also thankful to VIREO (Visual Resources for Ornithology, Academy of Natural Sciences, Drexel University) for allowing us to use photographs held in their repository.

We are grateful to Shannon Davies at Texas A&M University Press for her patience and willingness to shepherd us through the publishing process and to Noel Parsons for editing the final draft. Several individuals took time out of their schedules to provide extensive editorial reviews to early drafts of this book; their efforts are greatly appreciated. The published version of the book is a reflection of their suggestions.

Several generations of ornithologists and wildlife biologists have contributed to our understanding of Texas waterfowl through their monitoring efforts, research, and writings. Likewise, numerous observers have contributed sightings of rare and uncommon waterfowl to the Texas Bird Records Committee and the journal *North American Birds*; these contributions have allowed for the development of an extensive database related to the occurrence and distribution of rare Texas birds. In writing this book, we relied on the combined works of these individuals, and we acknowledge and appreciate their efforts.

Texas Waterfowl

Introduction

We hope this book will provide birders, hunters, naturalists, and others interested in learning about Texas waterfowl with a useful natural history resource. As we set out to write an account for each species of Texas waterfowl, we attempted to highlight the most interesting aspects of the bird's life history and, where possible, incorporate pertinent results from studies carried out in Texas. Comparable information was not available for all species, and the accounts vary in length and depth of coverage as a result. Longer accounts, those for common species and Texas breeders, start with an interesting fact about the species and then discuss Texas distribution, harvest, population status, diet, range, habitats, reproduction, and appearance. While there was not enough space to incorporate every aspect of the annual cycle (for example, we did not address molt migration, juvenile plumages, and the behavioral displays associated with courtship), we have included a source list at the end of each account and a full bibliography for those who would like to explore the literature further.

Although we describe plumages, this is not intended as a field or identification guide, as there are several excellent guides on the market. We used English measurements to strengthen appeal of the book to those outside the traditional science community. Finally, we restricted accounts to wild waterfowl; thus, we did not address domestic waterfowl, feral waterfowl (for example, semiwild park ducks), exotic species that likely escaped from captivity, or exotic species known to have been introduced to North America.

Common Terminology

dabbling duck: This group of ducks belongs to the genus *Anas.* They typically feed from the surface, and the term *dabbling* comes from their feeding methods. When foraging, they often submerge the front portion of their body in the water, and their rump tilts upward. This group includes many common species, such as Mallards.

diving duck: Sometimes referred to as bay ducks or pochards, this group of ducks belongs to the genus *Aythya.* They primarily forage under water by diving. This group includes Canvasbacks and Lesser Scaup. Other suites of ducks, such as sea ducks and stifftails, also forage by diving.

down: Down feathers are very small and fine and are found under the coarse outer feathers of waterfowl. Down serves to insulate waterfowl from inclement weather, and females of many species line their nest bowls with liberal amounts of down that they pluck from their breast. Incubating females cover their eggs with a layer of down, which they pull from the margins of the nest bowl, before leaving the nest unattended (that is, before taking an incubation recess). Down reduces the cooling rate of the eggs while the females are away. The small, fine feathers found on newly hatched waterfowl are also called down.

field feeding: Foraging in dry agricultural fields is known as field feeding. Geese and certain dabbling ducks, such as Mallards, Northern Pintails, Green-winged Teal, and American Wigeon are common field feeders.

fledge: Fledge, as it is used here, refers to the moment young waterfowl are capable of flight. For precocial young, such as ducklings, goslings, and cygnets, they are well developed and approximately adult-sized when they fledge.

incubation period: Although effective incubation (heat transfer from females to eggs that leads to embryo development) may begin well before the last egg is laid, traditionally the term *incubation period* was used to refer to the time period between the laying of the last egg and hatching. It is used in the traditional sense here.

longevity: As reported here, longevity records are measured in marked (banded) wild birds. The value is based on how old the bird was when it was banded and how much time elapsed between the banding date and the recovery date. For the recovery date to be on file, the band's unique number must be reported to the Bird Banding Laboratory in Laurel, Maryland. Bands are typically recovered by hunters. We address longevity only for birds with over 200,000 banding records, unless the record was Texas-specific.

nest parasitism: Nest parasitism is sometimes called brood parasitism or dump nesting. It occurs when a female lays one or more eggs in the nest of another female.

North American Waterfowl Management Plan: An overarching plan that guides waterfowl habitat conservation in Canada, the United States, and Mexico. The plan established population objectives for many species.

overwater nest: A nest that is constructed in the emergent, marshy vegetation zone of a wetland. Overwater nests are made by bending, breaking, and weaving emergent vegetation (for example, cattails) together to form a dense, matted mound or a floating platform. Shallow depressions, or nest bowls, are rounded out on top of platforms.

perching duck: This group of ducks frequently lands in trees, nests in cavities, and is associated with wooded areas. Although many other Texas waterfowl nest in cavities and occasionally perch in trees, Wood Ducks and Muscovy Ducks are the only North American members of this group. Species in this group do not necessarily share a common ancestry.

Prairie Pothole Region: This is a critically important breeding area for many North American ducks. Formerly prairie, the area now has a large agricultural component. It is pockmarked with abundant seasonal, semipermanent, and permanent wetlands. This area includes portions of Iowa, Minnesota, Nebraska, South Dakota, North Dakota, Montana, Manitoba, Saskatchewan, and Alberta. Throughout the text, wetlands in this region are referred to as prairie wetlands.

Prairie Parklands (or Canadian Parklands): Often combined with the Prairie Pothole Region (that is, as Prairie Pothole and Parkland Region), this region is interspersed with aspens, prairies, and numerous wetlands. It is a narrow ecological zone between the prairies and the boreal forests.

precocial: Refers to species that are relatively mature at the moment they hatch, in that they depart the nest soon after hatching, forage for themselves, and are able to maintain their own body temperature. All Texas waterfowl are largely precocial in that they maintain their body temperature with minimal brooding, leave the nest shortly after hatching, and are capable of foraging for themselves (although adults lead them to feeding areas).

sea duck: A loose group that includes eiders, scoters, Long-tailed Ducks, Harlequin Ducks, and other species. Sea ducks are largely marine in nature, although they may occur well inland. Most sea ducks dive when foraging. Many species have subadult plumages (distinct first-year patterns and sometimes distinct second-year patterns). Buffleheads, goldeneyes, and mergansers are often included in this group.

stiff-tailed duck: A small group of ducks that forage by diving and frequently hold their tails erect when sitting on the water. Ruddy Ducks and Masked Ducks are the only Texas waterfowl in the stifftail group.

Texas Mid-winter Waterfowl Survey: A cooperative state and federal waterfowl survey conducted annually during early January, although geese in some areas of the state may be counted in late December. Many other states also conduct a midwinter waterfowl survey.

upland nest: A nest that is constructed on the ground by scratching out a shallow depression (bowl) in soil or vegetation. Upland nests may be lined with liberal amounts of vegetation and down.

Treatment of Plumages

Many terms have been used to describe the plumages of waterfowl. *Eclipse*, *nonbreeding*, and *basic* have all been used to describe the duller, femalelike plumages worn during the post-breeding period (late summer, fall, and early winter). *Winter*, *breeding*, and *alternate* have been used to describe the brighter plumages worn when pair bonds are formed (this plumage often lasts through the nesting season). Although the terms *basic* and *alternate* tend to be used most often in literature, we used *breeding plumage* to refer to the brighter plumage and *nonbreeding plumage* to refer to the duller plumage. We address first-winter and subadult plumages only when they are distinctly different from the breeding plumage of adults. The timing of plumage acquisition and the duration that plumages are worn varies greatly among species; however, we do not address molt.

Texas Ecological Regions

We follow the ecological regions used in the *Texas Ornithological Society Handbook of Texas Birds.* These regions largely follow those developed by the Lyndon B. Johnson School of Public Affairs, University of Texas, Austin.

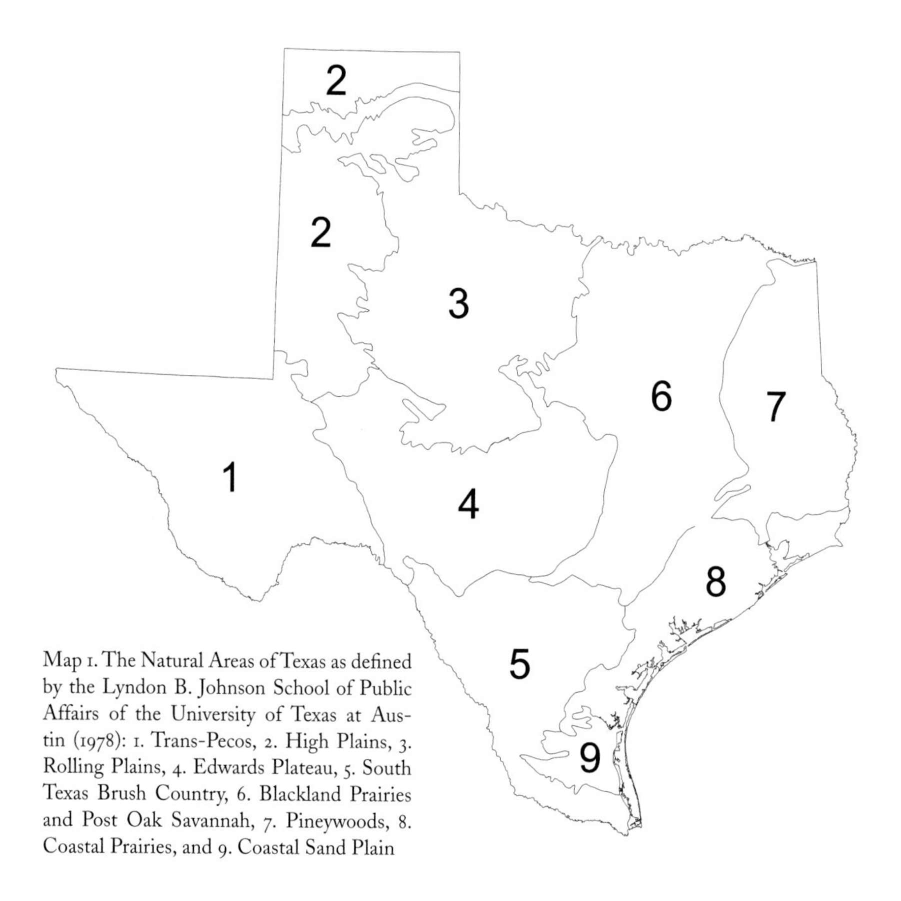

Map 1. The Natural Areas of Texas as defined by the Lyndon B. Johnson School of Public Affairs of the University of Texas at Austin (1978): 1. Trans-Pecos, 2. High Plains, 3. Rolling Plains, 4. Edwards Plateau, 5. South Texas Brush Country, 6. Blackland Prairies and Post Oak Savannah, 7. Pineywoods, 8. Coastal Prairies, and 9. Coastal Sand Plain

Abbreviations

NAWMP	North American Waterfowl Management Plan
NWR	National Wildlife Refuge
ppt	parts per thousand
TPWD	Texas Parks and Wildlife Department
USDA	US Department of Agriculture
USFWS	US Fish and Wildlife Service
USGS	US Geological Survey

Map Key

Year-round distribution

Winter distribution

Summer distribution

Migration

Irregular occurrence

Extent of irregular range

Used in the case of Texas Bird Record Committee Review Species, one dot signifies a single occurrence.

Species Accounts

Black-bellied Whistling-Duck (adult). *Photograph by Raymond S. Matlack, December 24, 2005, Brazos Bend State Park, Fort Bend County, Texas.*

BLACK-BELLIED WHISTLING-DUCK

Dendrocygna autumnalis

When a female lays one or more eggs in the nest of another female (either the same species or a different species), it is known as nest parasitism. This phenomenon is common among waterfowl, but Wood Ducks and Redheads are renowned for it. Black-bellied Whistling-Ducks, however, hold the record for what is perhaps the most extraordinary nest. At least 17 different females laid eggs in a single nest. This clutch topped out at 101 eggs. Amazingly, 38 eggs hatched.

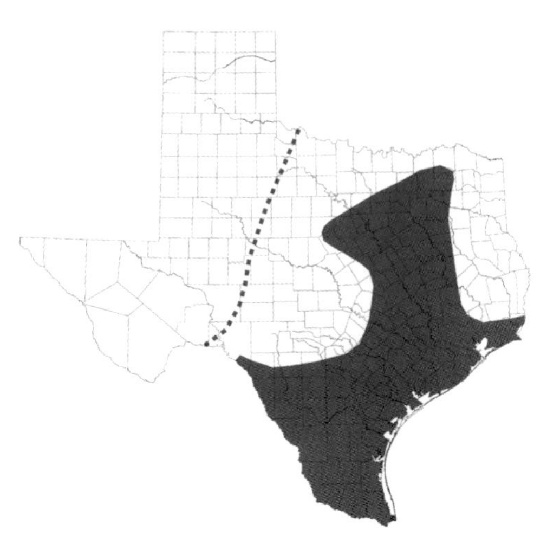

TEXAS DISTRIBUTION

Breeding: Texas is host to the largest number of breeding Black-bellied Whistling-Ducks in the United States. They are found in the South Texas Brush Country, Coastal Sand Plain, Coastal Prairies, Edwards Plateau, and Post Oak Savannah–Blackland Prairies. Occasional nesting occurs in the Rolling Plains and possibly other regions.

Migration and Winter: Most Black-bellied Whistling-Ducks depart during August, September, or October. However, some remain in the South Texas Brush Country during winter, and large roaming flocks occasionally occur in the Coastal Prairies. Spring migrants return to Texas in March and April.

TEXAS HARVEST

Harvest estimates in Texas averaged 4,849 annually from 1999 to 2006. This was about two-thirds of the US harvest.

LONGEVITY

The longevity record for a wild Black-bellied Whistling-Duck is eight years, two months. It was banded as an adult at Laguna Atascosa NWR in Cameron County and recaptured on a nest (during research) at the Welder Wildlife Foundation in San Patricio County.

POPULATION STATUS

Although Black-bellied Whistling-Ducks have long been present in South Texas, even hunted for markets, there is evidence to suggest that they largely disappeared during the early 1900s. However, they were again common in the Lower Rio Grande Valley by the 1950s. A northward expansion into Live Oak, San Patricio,

Kleberg, and Brooks Counties occurred by the early 1960s. Range expansions into the rice-growing regions of the state occurred after seed rice treated with aldrin (an organochlorine insecticide) was banned in the mid-1970s. They were well established throughout the central and upper portions of the Coastal Prairies by the mid-1980s. The Breeding Bird Survey suggests their abundance increased between 1966 and 2007, which corresponds with the northward expansion of their range in Texas. There is no population estimate for Black-bellied Whistling-Ducks.

DIET

The diet of Black-bellied Whistling-Ducks is mostly plant material. In Live Oak and San Patricio Counties, grain sorghum and bermuda grass were heavily consumed by breeding birds. In the Mexican state of Sinaloa, 97 percent of the diet of wintering birds was plant material; rice, corn, and wheat comprised about 75 percent of the foods consumed. Congregations of wintering birds often forage in agricultural fields.

RANGE AND HABITATS

Breeding: There are northern and southern subspecies of Black-bellied Whistling-Ducks. The southern subspecies breeds in the northern two-thirds of South America. The northern subspecies breeds in Florida, Louisiana, Texas, and Arizona and has recently been documented nesting as far north as Oklahoma and as far east as South Carolina. It also breeds along the east and west coasts of Mexico and in Central America. Black-bellied Whistling-Ducks readily use urban areas at all times of the year. Breeding pairs use stock ponds, abandoned gravel pits, shallow depressional wetlands, resacas, and lakes. They prefer shallow freshwater wetlands that contain floating plants (for example, water hyacinth), waterlilies, cattails, and dead trees. Wetlands they use often have thickets nearby. They nest in natural and artificial cavities (nest boxes). Upland nests are uncommon, although densities as high as six upland nests per acre may be found on islands. Upland nests may be located in herbaceous vegetation, on bare soil, and under prickly pear. Rarely, nests are found in chimneys and palm fronds.

Migration and Winter: Except for the northern limits of their range, their breeding and wintering ranges overlap. Likewise, wetlands used by migrating and wintering Black-bellied Whistling-Ducks are similar to those used by breeding ducks, although migrating and wintering birds use coastal wetlands, lagoons, mangrove swamps, and rivers to a greater degree. They prefer shallow water and often stand or walk while foraging. During the nonbreeding season they occasionally roam widely, showing up in places such as California and Ontario.

REPRODUCTION

Pair Bonds: Black-bellied Whistling-Ducks pair in their first winter. They usually form lifelong pair bonds but will re-pair if their mates are lost or die. Divorce (splitting of pair bonds) has been documented. Pairs commonly use the same nest cavity in successive years.

Nesting: In Texas, they nest from April through September. Competition for cavities may occur during the nesting season; there have been cases in which they have had their nest cavities usurped by Wood Ducks and Muscovy Ducks. Black-bellied Whistling-Ducks do not carry vegetation to the nest or add down. They lay eggs at a rate of 1 per day, and incubation begins in late laying. Their clutch size is approximately 13 eggs. Males and females incubate in alternating sessions that last approximately 24 hours. Their incubation period is about 28 days. Nest success for both cavity nests and upland nests is typically high. In a study that spanned the South Texas Brush Country, Coastal Sand Plain, and Coastal Prairies, nest success was approximately 30 percent in nest boxes. In Willacy County, success of ground nests was about 39 percent on islands. Most unsuccessful nesting attempts are a result of nest abandonment, loss of mates during incubation, and excessive nest parasitism. They renest at rates around 19 percent and will renest if their ducklings are lost. Up to three nesting attempts in one season and double brooding (two successful broods) have been documented. About 70 percent of nests are parasitized by other Black-bellied Whistling-Ducks. They have also been documented to lay eggs in nests of other species.

Ducklings: Adults lead ducklings permanently away from the nest 18–24 hours after they hatch. Both parents care for young. Frequently, one parent leads ducklings while the other follows closely behind. Adults brood ducklings sporadically during their first 12 days. Adults give vocal cues to lead the ducklings away from predators and will feign injury to lead predators away from ducklings. The young fledge at 53–63 days; adults often remain with their young even after they are capable of flight. Ducklings exposed to saline and hypersaline conditions likely have lowered survival, as they are highly susceptible to sodium toxicity.

APPEARANCE

Black-bellied Whistling-Ducks have a long neck and long legs, which gives them a gooselike appearance. They have one plumage year-round, and males and females are identical. They have a gray upper neck and head with a white eye ring. Their crown and hind neck are chestnut brown. The lower portions of their neck, breast, and back are rufous (rust-colored). Their belly, rump, and tail are black. Their folded wings appear olive to off-white. Their bill is reddish with a bluish tip. Adults have pink legs. During summer, females and males weigh approximately 1.9 and 1.8 pounds, respectively.

SOURCES

INTRODUCTION: Delnicki et al. 1976. TEXAS DISTRIBUTION: Bolen et al. 1964; Lobpries 1987; Lockwood and Freeman 2004. TEXAS HARVEST: Kruse 2007. LONGEVITY: Delnicki 1973; Clapp et al. 1982. POPULATION STATUS: Lobpries 1987; James and Thompson 2001; Brush 2005; Sauer et al. 2008. DIET: Bolen and Forsyth 1967; Kramer and Euliss 1986. RANGE AND HABITATS: Meanley and Meanley 1958; Bolen et al. 1964; Johnsgard 1978; Bellrose 1980; Bolen and Rylander 1983; Markum and Baldassarre 1989; Bergstrom 1999; James and Thompson 2001; Kamp and Loyd 2001; Harrigal and Cely 2004; Brush 2005; Edmonds and Stolley 2008. REPRODUCTION: Bolen et al. 1964; Bolen 1967, 1971a, 1971b; Bolen and Cain 1968; Cain 1968; Delnicki et al. 1976; Bolen and McCamant 1977; Bolen and Smith 1979; McCamant and Bolen 1979; Bellrose 1980; Bolen and Rylander 1983; Delnicki 1983; Heins 1984; O'Kelley 1987; Markum and Baldassarre 1989; James and Thompson 2001; Edmonds and Stolley 2008; Stolley et al. 2008. APPEARANCE: Bolen 1964; Bellrose 1980; James and Thompson 2001.

FULVOUS WHISTLING-DUCK

Dendrocygna bicolor

No species of waterfowl is as closely associated with agriculture—moreover, a particular crop—as Fulvous Whistling-Ducks are to rice production. Wherever rice is grown within their range, they use it heavily. They forage, nest, and raise their young in rice fields. There is even evidence that they prefer to nest in rice fields over natural habitats. Additionally, recent breeding range expansions are associated with their further colonization of rice-growing regions.

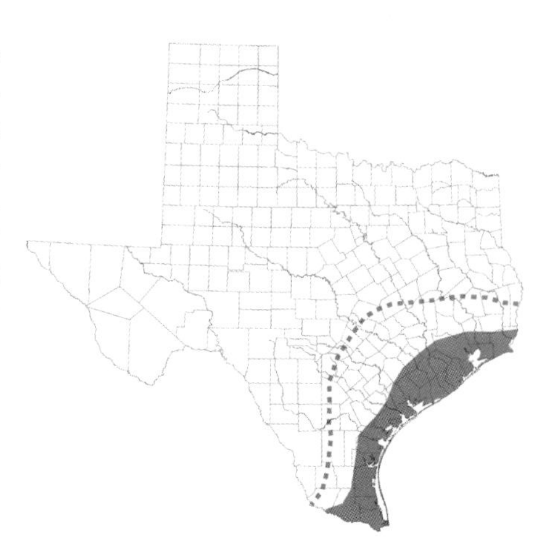

TEXAS DISTRIBUTION

Breeding: Fulvous Whistling-Ducks breed in the South Texas Brush Country, Coastal Sand Plain, and Coastal Prairies. Nesting has also been documented in the Post Oak Savannah–Blackland Prairies.
Migration and Winter: The Texas population is migratory; most are gone by November. They are rare to locally common during winter within their breeding range. Spring migrants return in early to mid-March.

TEXAS HARVEST

From 1999 to 2006, estimated US harvest (including Texas) of Fulvous Whistling-Ducks was less than one thousand annually.

POPULATION STATUS

There is no population estimate for Fulvous Whistling-Ducks. The Breeding Bird Survey suggests their abundance was stable to increasing from 1980 to 2007. Prior to that, their population experienced declines, which were likely associated with the use of aldrin-treated seed rice.

DIET

Seeds accounted for over 96 percent of the diet of adult Fulvous Whistling-Ducks collected in Louisiana rice fields during the nesting season; rice consumption was low (< 4 percent) early in the breeding season but increased (25 percent) during incubation. The most common seeds consumed were flatsedge, signalgrass, and beaksedge. Fulvous Whistling-Ducks are perceived as a nuisance because they occasionally use rice fields in large numbers during spring, about the time rice is planted. However, they consume seeds of undesirable plants (weeds), including "red rice," during spring.

Fulvous Whistling-Duck (adult). *Photograph by Greg Lasley, March 1996, Brazos Bend State Park, Fort Bend County, Texas.*

RANGE AND HABITATS

Breeding: Fulvous Whistling-Ducks have an amazing distribution that includes North, Central, and South America; the Caribbean, Hawaii, Madagascar, India, and Sri Lanka; and portions of Africa and Burma. In the continental United States they primarily breed in California, Texas, Louisiana, and Florida. Breeding pairs in the United States use rice fields, shallow freshwater wetlands, freshwater marshes, and temporarily flooded pastures and prairies. They make upland nests and overwater nests. Most overwater nests are found in rice fields.

Migration and Winter: Although they are resident throughout much of the Americas, most Fulvous Whistling-Ducks that breed in the continental United States migrate to more southerly areas during winter. Migrants use rice fields and shallow freshwater marshes, particularly those with floating plants. In Mexico, wintering Fulvous Whistling-Ducks use river deltas, mangrove forests, brackish lagoons, freshwater marshes, lakes, and flooded grasslands.

REPRODUCTION

Pair Bonds: Solid information is not available concerning pairing chronology or the type of pair bonds formed (that is, seasonal or long-term).

Nesting: In Louisiana, Fulvous Whistling-Ducks nest from April through August, but in South America nesting is concurrent with rainy seasons. They may line nest bowls with vegetation that comes from beyond the immediate vicinity of the nest. They do not add down to the nest. They lay eggs at a rate of 1 per day, and their clutch size is 9–14 eggs. Both males and females incubate. Their incubation period is 24–25 days. Upland nesting pairs spend about 97 percent of their time incubating, and overwater nesting pairs spend about 89 percent of their time incubating. In San Patricio County, over 50 percent of nests located in nonagricultural habitats were successful. Renesting is likely. They frequently parasitize the nests of other Fulvous Whistling-Ducks.

Ducklings: Hatching is largely synchronous, and ducklings permanently depart the nest the morning after they hatch. Both parents brood ducklings after they leave the nest. Adults will feign injury in order to lead predators away from ducklings, and they regularly lead ducklings overland, particularly in response to dewatering of rice fields. Ducklings are capable of flight at about 63 days.

APPEARANCE

The long neck and legs of Fulvous Whistling-Ducks give them a gooselike appearance. Males and females are identical and have one plumage year-round. At a distance they have a tawny brown appearance that is darker above than below. They have a fulvous-colored head, breast, and belly. Their chin is pale white. Their back is brown with fulvous barring, and their tail and wings are blackish brown. They have buffy yellow stripes across their flanks. Their bill is slaty gray, and they have blue legs. Their weight fluctuates seasonally. During the breeding season males and females average 1.7 and 1.6 pounds, respectively.

SOURCES

INTRODUCTION: Carroll 1932; Bolen and Rylander 1983; Peris et al. 1998; Hohman and Lee 2001. TEXAS DISTRIBUTION: Flickinger et al. 1977; Benson and Arnold 2003; Lockwood and Freeman 2004. TEXAS HARVEST: Kruse 2007. POPULATION STATUS: Flickinger et al. 1977; Lobpries 1987; Peris et al. 1998; Hohman and Lee 2001; Sauer et al. 2008. DIET: Mugica Valdes 1993; Hohman et al. 1996; Hohman and Lee 2001. RANGE AND HABITATS: Meanley and Meanley 1959; Zwank et al. 1988; Hohman and Lee 2001. REPRODUCTION: Cottam and Glazener 1959; Meanley and Meanley 1959; Flickinger 1975; Hohman and Lee 2001; Pierluissi 2006. APPEARANCE: Johnsgard 1975; Gooders and Boyer 1986; Hohman and Lee 2001.

GREATER WHITE-FRONTED GOOSE

Anser albifrons

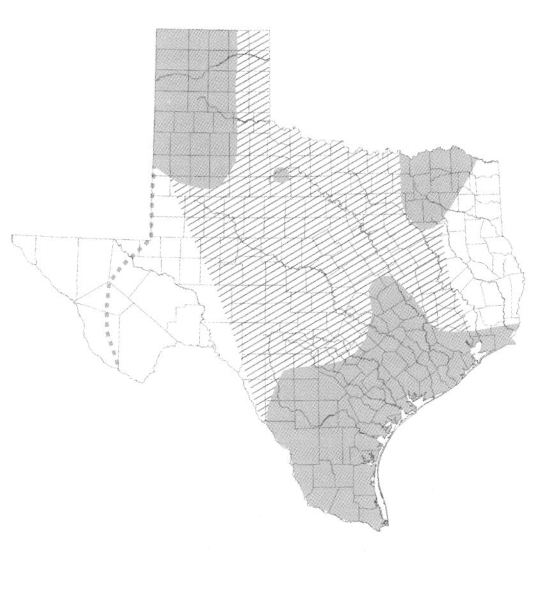

Parents and young of most North American geese remain together during their first winter and into spring. However, in Greater White-fronted Geese, family members may continue to associate with one another well beyond their first winter. In California and Oregon, 39 percent of two-year-old geese (second winter) and 38 percent of three-year-old and older geese continued to associate with their parents during winter. Interestingly, two-year-old and three-year-old siblings also associated with each other on occasion, even in the absence of parents.

TEXAS DISTRIBUTION

Breeding: Greater White-fronted Geese do not breed in Texas.

Migration: Greater White-fronted Geese are common in the central portion of Texas during migration. They are rare to uncommon in the remainder of the state. Fall migrants passing through the Rolling Plains and Post Oak Savannah–Blackland Prairies peak in late October or November, and spring migrants peak in February or early March. Statewide, fall migrants typically appear in late September, and spring migrants may linger into early April.

Winter: Traditionally, they wintered almost exclusively in coastal marshes, although they are now commonly found inland. From 2001 to 2008 they averaged 220,000 during the Texas Mid-winter Waterfowl Survey (TPWD unpublished). They are most common in the Coastal Prairies and South Texas Brush Country (TPWD unpublished), and they are locally abundant in the Rolling Plains and Post Oak Savannah–Blackland Prairies. They are rare to uncommon in the rest of the state.

TEXAS HARVEST

From 1999 to 2006, harvest of Greater White-fronted Geese in Texas averaged 88,291 annually. This was about 36 percent of their annual US harvest.

POPULATION STATUS

The North American Waterfowl Management Plan's population goal for Greater White-fronted Geese is 910,000. The goal for the population that winters in Texas, the Midcontinent Population, is 600,000. In 2011 this population averaged 709,800, and it has a stable trend.

DIET

Greater White-fronted Geese consume vegetation year-round. Plants consumed in Alaska during the breeding season include arrow grass (bulbs) and pendant grass (shoots, roots). Berries, seeds, and sedges are also consumed during the breeding season. In Sinaloa, common foods included alkali-bulrush, barnyard grass, soybeans, and wheat. Important foods in Texas likely include winter wheat, sorghum, corn, and rice. In Nebraska, corn and winter wheat accounted for 90 percent and 9 percent, respectively, of the foods consumed during spring migration.

RANGE AND HABITATS

Breeding: There are three subspecies of Greater White-fronted Geese in North America. Greenland White-fronted Geese (*A. a. flavirostris*) breed in western Greenland and winter in Ireland and Britain. Tule Geese (*A. a. gambeli*), which are larger and darker than Greater White-fronted Geese (*A. a. frontalis*), breed near Cook Inlet, Alaska, and winter in the Sacramento Valley and the Sacramento–San Joaquin Delta, California. Greater White-fronted Geese breed in inland tundra and taiga regions of northeast Asia, Alaska, and northwestern Canada. Breeding pairs are associated with freshwater arctic wetlands, mudflats, meltwater pools, meadows, bogs, taiga forests, streams, and rivers. They nest on the ground. Nest locations are typically on sedge- or grass-covered hummocks near the base of trees (for example, spruces), among shrubs (for example, raspberries), on grassy or brushy slopes, or at the base of rocky outcrops. Nests may be well concealed or relatively open.
Migration: Greater White-fronted Geese congregate in arctic estuaries, brackish marshes, and river deltas prior to fall migration and upon their return to the arctic in spring. Wetlands in southeastern Alberta and southwestern Saskatchewan are critically important to migrants, as are Nebraska's rainwater basins. Migrants use prairie rivers, wet meadows, post-harvest grain fields (for example, corn fields), and winter wheat fields. In Texas, shallow depressional wetlands, reservoirs, and stock ponds are also used.
Winter: During winter, Greater White-fronted Geese are managed as two popu-

Greater White-fronted Goose (adult). *Photograph by Raymond S. Matlack, December 30, 2008, Canyon, Randall County, Texas.*

lations: the Pacific and Midcontinent Populations. The Pacific Population winters sporadically from southwestern British Columbia to Colima, Mexico. The Midcontinent Population winters in Arkansas, Louisiana, Texas, and eastern Mexico. Greater White-fronted Geese have the widest range of any goose in Mexico. Throughout their winter range they use post-harvest grain fields, fallow fields, flooded and dry rice fields, wet meadows, shallow depressional wetlands, tidal mudflats, shallow open wetlands, river deltas, coastal estuaries, marshes, and bays. In Texas they are frequently associated with stock ponds, improved pastures, and disked soybean and sorghum fields (particularly in the Coastal Prairies).

REPRODUCTION

Pair Bonds: Greater White-fronted Geese are thought to pair after their second summer. They mate for life but will form new pair bonds if their mate is lost. Males follow their mates north in spring. Females have a high degree of fidelity to breeding areas.

Nesting: Related females tend to nest in close proximity to each other. Females select nest sites and construct nests. They line nest bowls with vegetation obtained from the vicinity of the nest site. They add down to the nest about the time the last egg is laid. Females lay about one egg every 30–32 hours, and their clutch size is four or five eggs. Only females incubate; they spend over 95 percent of their time on the nest, and their incubation period is 24–25 days. Males remain near their mates during egg laying and incubation and join them during incubation recesses. Offspring from the previous year are often still bonded with their parents and remain with them at the nest territory at least through late incubation. Renesting is doubtful if clutches are lost during incubation, and they will not renest if their goslings are lost.

Goslings: Parents lead goslings permanently away from the nest about 24–48 hours after they hatch. Both parents attend to the young. Females brood goslings in the nest and continue to brood them after departure, particularly during inclement weather. Family groups, consisting of parents and goslings, often aggregate with other family groups while rearing their young. All adults in these larger family groups may help defend goslings from threats. Goslings fledge at 42–49 days and migrate south with their parents.

APPEARANCE

Adults (*year-round*): Males and females are similar in appearance. They have a white area on the forward portion of their head, near the bill. Their body is mostly brown, although they have heavy, dark barring (or speckling) across their lower breast and belly. The dark brown feathers on their back have light brown edging. When they are swimming or standing, a thin, white horizontal line shows along their sides. The area under their tail is large and white. Their bill is pinkish. The weight of adults varies throughout the annual cycle. During fall, adult males and females in Saskatchewan weighed about 6.0 and 5.4 pounds, respectively.

First winter: In early fall, plumage of first-winter Greater White-fronted Geese is similar to that of adults, except that young geese lack heavy barring on their breast and belly. They also do not have white on their head. Both the amount of white on their head and the amount of dark barring on their belly increase during winter.

SOURCES

INTRODUCTION: Bellrose 1980; Ely 1993. TEXAS DISTRIBUTION: Pulich 1988; Ballard and Tacha 1995; Anderson and Haukos 2003; Lockwood and Freeman 2004; USFWS 2008a. TEXAS HARVEST: Kruse 2007. POPULATION STATUS: NAWMP 2004; Warner et al. 2007; USFWS 2011. DIET: Smith et al. 1989; Budeau et al. 1991; Migoya and Baldassarre 1993; Ely and Dzubin 1994; Krapu et al. 1995. RANGE AND HABITATS: Bellrose 1980; Ely and Raveling 1984; Leslie and Chabreck 1984; Smith et al. 1989; Warren et al. 1992; Ely and Dzubin 1994; Ballard and Tacha 1995; Krapu et al. 1995; Carrière et al. 1999. REPRODUCTION: Bellrose 1980; Ely and Raveling 1984; Batt et al. 1992; Warren et al. 1992; Ely 1993; Ely and Dzubin 1994; Fowler et al. 2004. APPEARANCE: Bellrose 1980; Ely and Dzubin 1994.

SNOW GOOSE

Chen caerulescens

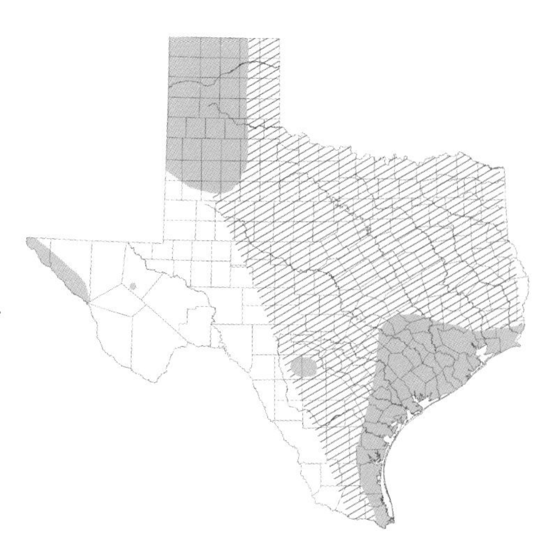

It is common for young waterfowl to be led overland by adults. This may be due to drying wetlands or deteriorating food resources. It is also common for the young of many species to travel several miles before they are capable of flight. Adult Snow Geese and their goslings, however, are continuously on the move. Goslings have been documented to walk over 37 miles before they were capable of flight.

TEXAS DISTRIBUTION

Breeding: Snow Geese do not breed in Texas.
Migration: Migrant Snow Geese potentially occur throughout the state, but they are most common in the eastern two-thirds. Migrants appear in early October and linger into early April.
Winter: Traditionally, Snow Geese wintered almost exclusively in coastal marshes. Their winter range did not expand into the agricultural regions of the Coastal Prairies until after the 1920s. They now regularly occur throughout the Coastal Prairies, in the western Trans-Pecos, and in the High Plains. They are locally common in the Rolling Plains, South Texas Brush Country, Coastal Sand Plain, and Post Oak Savannah–Blackland Prairies, but they may occur anywhere in the state. From 2001 to 2008 the estimate for light geese (that is, combined Ross's and Snow Geese) during the Texas Mid-winter Waterfowl Survey was 804,500 (TPWD unpublished). In the High Plains and Coastal Prairies, Snow Geese comprise about 77 percent and 94 percent of light geese, respectively.

TEXAS HARVEST

From 1999 to 2006, regular-season Snow Goose harvest in Texas averaged 183,706 annually, which was about one-third of the annual harvest in the United States.

LONGEVITY

The longevity record for a wild Snow Goose is 27 years, six months.

POPULATION STATUS

The North American Waterfowl Management Plan's goals for the Midcontinent Population and Western Central Flyway Population are 1 million–1.5 million and 110,000, respectively. These are the two populations that winter in Texas. Since 1969, both of these populations have increased by over 300 percent. In 2011 the Midcontinent

Snow Goose (adult). *Photograph by Raymond S. Matlack, January 18, 2008, San Bernard National Wildlife Refuge, Brazoria County, Texas.*

Population was 3.2 million, and the Western Central Flyway Population was 196,100. The population explosion is due in part to their expansion into new wintering grounds and their exploitation of agriculture-based habitats. Population levels are no longer limited by poor body condition or high mortality rates during winter.

DIET

Plants commonly consumed during the breeding season include alpine chickweed, cranberry, horsetail, sedges, and mosses. During January, about 89 percent of the foods consumed in brackish marshes of Louisiana were belowground parts of marshhay cordgrass, saltmarsh bulrush, Olney bulrush, and saltgrass. In the Coastal Prairies, Snow Geese foraging in rice fields consumed rice grains in fall through early winter. In midwinter their diet shifted to aboveground parts of grasses and forbs. In fallow rice fields, about two-thirds of their diet was weed seeds (other than rice). Spikerush, forbs, and grasses accounted for 99 percent of their diet in pastures along the Texas coast. During spring, migrants in agricultural regions of North America frequently consumed corn, soybeans, and winter wheat.

RANGE AND HABITATS

Breeding: Snow Geese breed in numerous scattered colonies across the Canadian arctic, Alaska, northeast Siberia (Wrangel Island), and Greenland. In ornithologi-

cal literature, Snow Geese are often split into Lesser (*C. c. caerulescens*) and Greater (*C. c. atlantica*) subspecies. The two subspecies differ in their breeding and wintering regions (Greaters breed in the eastern arctic and winter on the eastern seaboard of North America) and size. Snow Geese nest on the ground. Nests are found on islands, slight knolls, rolling hills, ridges, and hummocks and near shallow lakes and streams. They occasionally nest on slopes of ravines and cliff edges. Vegetation used for nesting includes sedges, dwarf birch, arctic bell heather, and mountain cranberry.
Migration: Migrants use brackish estuaries, river deltas, wet meadows, postharvest grain fields (for example, corn fields), winter wheat, large prairie wetlands, playa wetlands, rainwater basins, and prairie rivers.
Winter: Snow Geese are managed and tracked based on their wintering ground affiliations. The Western Population winters in scattered locations from southeastern British Columbia to Baja California and Sonora. The Western Central Flyway Population winters in western Texas, western Oklahoma, western Kansas, eastern Colorado, and New Mexico. This population also winters in Mexico's northern highlands. The Midcontinent Population winters along the Gulf coast from Veracruz to Louisiana. This population has recently expanded its winter range northward into Nebraska, Iowa, and western Illinois. The Eastern Population winters along the Atlantic coast from Massachusetts to South Carolina and consists mostly of Greater Snow Geese. Wintering Snow Geese use winter wheat, postharvest grain fields, fallow fields, rice fields, wet meadows, playa wetlands, coastal marshes, river deltas, tidal

Snow Goose (adult, dark morph, often called blue goose). *Photograph by Mark W. Lockwood, January*

mudflats, industrial impoundments, and shallow open wetlands. In portions of the Coastal Prairies they often use disked sorghum fields.

REPRODUCTION

Pair Bonds: Pair bond formation occurs during their second or third winter. They pair for life, although they will form new pair bonds if their mate is lost. Snow Geese show a high degree of fidelity to both breeding and wintering areas. Snow Geese from mixed colonies of blue- and white-morph birds tend to choose mates that reflect the color morph of their family group (mixed, white, or blue).

Nesting: Snow Geese primarily nest in colonies. Females construct nests as soon as snow clears from nesting areas. Nest bowls are lined with down and vegetation. Vegetation comes from the vicinity of the nest. Nest sites are frequently reused from year to year. Females lay about one egg every 33 hours. Their clutch size is about four eggs. Once females commence incubation, males may leave temporarily to pursue mating opportunities with other females. Both males and females defend the nest site. Only females incubate; they take multiple incubation recesses per day. Their incubation period is about 24 days. Renesting may occur if nests are lost early during the egg laying period. Renesting is doubtful if nests are lost during incubation, and they will not renest if their goslings are lost. Females commonly parasitize the nests of other Snow Geese.

Goslings: Parents lead goslings permanently away from the nest about 24 hours after they hatch. Females brood young occasionally. Males will attempt to drive predators away from goslings. Family groups, consisting of parents and goslings, often form loose associations with other family groups. Goslings are capable of flight at around 43 days. They migrate south with their parents. Young and parents typically remain together through spring, sometimes even into the next breeding season.

APPEARANCE

Adults (*year-round*): There are two color morphs of Snow Geese, white and blue. At one time the two color morphs were considered separate species. Within each color morph, males and females share identical plumages and are similar in appearance year-round. White-morph geese are white except for blackish primaries (wing tips). Blue-morph geese, which are commonly called blue geese, come in several plumage patterns. These patterns grade from dark coverage over their entire body, except for the head, to dark only on their lower neck, back, flanks, and wings. The dark pattern of blue-morph geese is a mix of gray, brown, black, and bluish feathers, and it tends to have a frosted appearance. Traditionally, blue-morph Snow Geese had very localized summer and winter ranges, but both their breeding and wintering distributions are now more widespread. Both color morphs have dark eyes and a pinkish bill. On the Coastal Prairies, wintering males and females averaged 4.9 and 4.4 pounds, respectively.

First winter: The first-winter plumage of white-morph Snow Geese has a grayish, dirty appearance. The first-winter plumage of blue-morph Snow Geese is browner than the plumage of adults. They also have a dark, slaty gray head and neck.

SOURCES

INTRODUCTION: Bellrose 1980; Slattery and Alisauskas 2007. TEXAS DISTRIBUTION: Hobaugh 1984; Ballard and Tacha 1995; Robertson and Slack 1995; Lockwood and Freeman 2004; USFWS 2008. TEXAS HARVEST: Kruse 2007. LONGEVITY: Lutmerding and Love 2011. POPULATION STATUS: Cooke et al. 1995; Abraham et al. 1996; Batt 1997; Mowbray et al. 2000; NAWMP 2004; Kruse et al. 2007; USFWS 2011. DIET: Hobaugh 1984; Alisauskas et al. 1988; Alisauskas and Ankney 1992; Gauthier 1993b; Miller et al. 1996; Gloutney et al. 2001. RANGE AND HABITATS: Batt 1977; Bellrose 1980; McLaren and McLaren 1982; McLandress 1983; Alisauskas and Ankney 1992; Turner et al. 1994; Tremblay et al. 1997; Alisauskas 1998; Elphick and Oring 1998; Mowbray et al. 2000; Vrtiska and Sullivan 2009. REPRODUCTION: Cooke and McNally 1975; Ankney and MacInnes 1978; Bellrose 1980; Prevett and MacInnes 1980; Lank et al. 1989; Cooch et al. 1991a, 1991b, 1993; Schubert and Cooke 1993; Williams et al. 1993, 2008; Cooke et al. 1995; Mowbray et al. 2000. APPEARANCE: Ankney 1982; Mowbray et al. 2000; Haukos et al. 2001.

ROSS'S GOOSE

Chen rossii

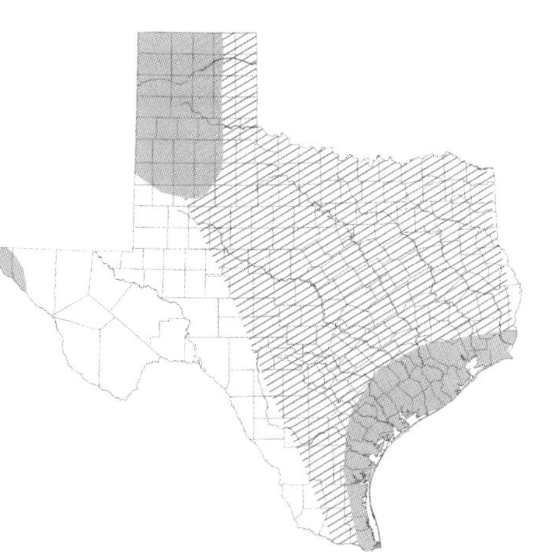

In his 1963 *A Field Guide to the Birds of Texas and Adjacent States*, Roger Tory Peterson described Ross's Goose as a casual visitor to Texas and reported sightings from only 5 counties. Today these geese are widespread in the state, wintering in more than 60 counties. They also regularly occur in the central third of the state during migration.

TEXAS DISTRIBUTION

Breeding: Ross's Geese do not breed in Texas.

Migration: During migration, they are generally found with flocks of Snow Geese. Migrants potentially occur throughout Texas but are most common in the central third of the state.

Winter: Wintering Ross's Geese are most common in the El Paso area, High Plains, and Coastal Prairies. Due to their similarities with Snow Geese, the two species are not differentiated during the Texas Mid-winter Waterfowl Survey. In the High Plains they represent about 23 percent of the light geese (that is, combined Ross's and Snow Geese). The first High Plains records were from Randall County in 1966. In the Coastal Prairies, they represent about 6 percent of light geese. The first coastal records were from Jefferson County in 1953 and Colorado and Wharton Counties in 1954.

TEXAS HARVEST

From 1999 to 2006, regular-season Ross's Goose harvest in Texas averaged 30,295 annually. This was about 48 percent of their annual harvest in the United States.

Ross's Goose (adult). *Photograph by Mark W. Lockwood, November 28, 2009, Andrews, Andrews County, Texas.*

POPULATION STATUS

Both their breeding and wintering range have greatly expanded. The North American Waterfowl Management Plan's population goal for Ross's Geese is 100,000. From 2001 to 2003 their annual population averaged 619,000. Considering that their population may have numbered less than 10,000 during the first half of the 20th century, the current population size is phenomenal.

DIET

On the breeding grounds they forage on mosses, sedges, leaves, rhizomes, and roots. During winter they forage on alfalfa, rice (seeds), saltgrass, winter wheat, and waste grains (for example, corn).

RANGE AND HABITATS

Breeding: Most Ross's Geese (95 percent) breed in Nunavut's Queen Maude Gulf.

They also breed on Baffin Island (Nunavut), around Hudson Bay, and locally in Alaska. Breeding pairs are primarily associated with wet meadows and marshy tundra. They nest on the ground and frequently use islands. Nests are found among sedges, heath tundra, and dwarf birches and in open, gravelly areas characterized by mosses, sparse vegetation, and large rocks. Nests are often situated so that they are protected from high winds.
Migration: Migration is largely through interior North America. Habitats used during migration include wet meadows, playa wetlands, large open wetlands, postharvest grain fields (for example, corn fields), and winter wheat.
Winter: There are three recognized wintering populations. The Pacific Flyway Population winters in California. The Western Central Flyway Population winters from Zacatecas, Mexico, north to Colorado. This population also winters in the Oklahoma and Texas Panhandles. The Midcontinent Population winters along the Texas and Louisiana coasts but may be found inland to Arkansas. Sightings outside these major wintering areas are not uncommon. Wintering Ross's Geese are frequently associated with flocks of Snow Geese. They use wet meadows, coastal marshes, playa wetlands, open shallow wetlands, winter wheat fields, postharvest grain fields, and both flooded and dry rice fields.

REPRODUCTION

Pair Bonds: Ross's Geese pair for life but probably form new pair bonds if their mate is lost. Age of pair bond formation in the wild is unknown, but captives perform their first courtship displays at about 19 months. They have a high degree of fidelity to both breeding and wintering areas, although some mixing between wintering populations is likely.
Nesting: Ross's Geese nest in colonies that may approach 146 nests per acre. Nest construction is an activity of the female, which lines its nest with plant material and down. Neither males nor females leave the nesting territory during prelaying and laying. Females lay about one egg every 29 hours. Their clutch size is typically four eggs. Only females incubate. After the last egg is laid, females spend approximately 87 percent of their time on the nest. Females rarely wander from the nesting territory during incubation breaks. While females are incubating, males stay nearby; their role is to defend the eggs and female. The incubation period lasts 21–22 days. Renesting has not been documented.
Goslings: Shortly after the eggs hatch, parents lead goslings permanently away from the nest. Parents are vigilant for predators and may even attempt to drive them away. Goslings can likely fly by 40–45 days. Family groups, consisting of parents and young, typically remain together during fall migration and winter and sometimes even into spring.

APPEARANCE

Adult (*year-round*): Males and females have identical plumage and are similar in appearance year-round. They are white except for blackish primaries (wing tips). The front and sides of their neck have a furrowed appearance. Their eyes are dark and their bill is pinkish. They have numerous bumpy, warty structures along the base of their bill. Compared to Lesser Snow Geese, their neck appears short, and their bill is small relative to their head. Blue-morph Ross's Geese occur but are extremely rare (far less than 0.01 percent). Many authorities believe all blue-morph Ross's Geese may be the result of past or recent hybridization with blue-morph Snow Geese. Ross's Geese reach their peak weight during egg laying; during this time both males and females weigh about 3.6 pounds.

First winter: Plumage of first-winter Ross's Geese is similar to that of adults but is tinged in gray. However, first-winter Ross's Geese are not as gray as first-winter Snow Geese.

SOURCES

INTRODUCTION: Peterson 1963; Lockwood and Freeman 2004. TEXAS DISTRIBUTION: Miller 1954; Buller 1955; Moser 2001; Seyffert 2001; Lockwood and Freeman 2004; USFWS 2008; Thorpe 2009. TEXAS HARVEST: Kruse 2007. POPULATION STATUS: Dzubin 1965; Ryder and Alisauskas 1995; NAWMP 2004; Kruse et al. 2007. DIET: Ryder and Alisauskas 1995; Elphick and Oring 1998; Glountney et al. 2001; Drewien et al. 2003. RANGE AND HABITATS: Bellrose 1980; McLandress 1983; Kerbes 1994; Turner et al. 1994; Ryder and Alisauskas 1995. REPRODUCTION: Ryder 1967, 1972; Palmer 1976a; Bellrose 1980; Melinchuk and Ryder 1980; Slattery 1994; Turner et al. 1994. APPEARANCE: McLandress and McLandress 1979; Ryder and Alisauskas 1995; Moser 2001.

CANADA GOOSE

Branta canadensis

CACKLING GOOSE

Branta hutchinsii

Most authorities traditionally recognized 11 subspecies of Canada Geese. However, in 2004 the American Ornithologists' Union recommended splitting Canada Geese into two distinct species. The two new species are Cackling Geese and Canada Geese. There are still multiple subspecies, which are aligned with either Canada Geese or Cackling Geese based on genetic, geographic, and behavioral similarities. Cackling Geese consist of 5 "small- bodied" subspecies: Bering (*B. h. asiatica*), Aleutian (*B. h. leucopareia*), Richardson's (*B. h. hutchinsii*), Ridgway's (*B. h. minima*), and Taverner's (*B. h. taverneri*). Canada Geese consist of 7 "large-bodied" subspecies: Atlantic (*B. c. canadensis*), Dusky (*B. c. occidentalis*), Giant (*B. c. maxima*), Great Basin (*B. c. moffitti*), Hudson Bay (*B. c. interior*), Lesser (*B. c. parvipes*), and Vancouver (*B. c. fulva*). For research conducted prior to 2004, it is often difficult to determine which species was studied. Because of this, the two species are treated together in this account, and the term *white-cheeked goose* is used for both species.

TEXAS DISTRIBUTION

Breeding: Cackling Geese do not breed in Texas, but Canada Geese are increasingly common breeders in the High Plains, Rolling Plains, and Post Oak Savannah–Blackland Prairies. They may also breed in other areas of the state. Texas breeders

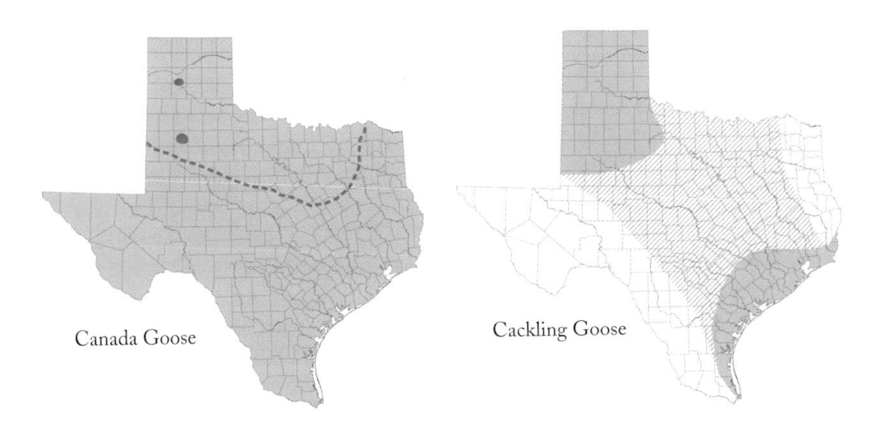

may be the result of intentional introductions, relocations of nuisance geese, or dispersal from stocking efforts in adjacent states. Small numbers were released at Hagerman NWR, Grayson County, in the late 1940s, and about 200 were stocked in the Athens area from 1985 to 1990. Approximately 9,000 were stocked in Oklahoma during the 1980s; in 1989, Oklahoma's estimated resident population was 21,000. *Migration:* White-cheeked geese may occur throughout the state during migration but are typically only common in the High and Rolling Plains. They are locally

Canada Goose. *Photograph by Raymond S. Matlack, November 18, 2005, Canyon, Randall County, Texas.*

Cackling Goose. *Photograph by Mark W. Lockwood, January 29, 2005, Lubbock, Lubbock County, Texas.*

common in the Post Oak Savannah–Blackland Prairies and Coastal Prairies. They arrive in Texas in late September and are notable by late October. In the High Plains and Coastal Prairies, their numbers peak in mid-January. Most nonresident white-cheeked geese depart by March.

Winter: Wintering white-cheeked geese may occur throughout the state. They are common in the High and Rolling Plains and locally common in the Coastal Prairies and Post Oak Savannah–Blackland Prairies. From 2001 to 2008, white-cheeked geese averaged 219,910 during the Texas Mid-winter Waterfowl Survey.

Richardson's is the primary Cackling Goose subspecies in Texas. It occurs in northwestern Texas and along the coast. In contrast, several subspecies of Canada Geese regularly winter in Texas, including Great Basin, Giant, Hudson Bay, and Lesser. Lessers are common in northwestern Texas. In flocks where Richardson's Cackling Geese and Lesser Canada Geese co-occur, the two species tend to segregate.

TEXAS HARVEST

From 1999 to 2006, white-cheeked goose harvest in Texas averaged 83,737 annually. This was about 3 percent of the annual white-cheeked goose harvest in the United States.

LONGEVITY

The longevity record for a wild white-cheeked goose is 33 years, three months.

POPULATION STATUS

The three populations of white-cheeked geese that winter in Texas include Short Grass Prairie, Tall Grass Prairie, and Western Prairie Populations. The North American Waterfowl Management Plan's objective for these populations is 417,000. In 2011 the combined annual index for Short Grass Prairie and Tall Grass Prairie Populations was over 700,000. These two populations comprise the vast majority of Texas birds and primarily consist of Lesser Canada Geese and Richardson's Cackling Geese. Wintering geese in northeast Texas are part of the Western Prairie Population, which comprises Great Basin, Giant, and Hudson Bay Canada Geese. Continent-wide, white-cheeked geese are well above the North American Waterfowl Management Plan's objective of 3,476,000.

DIET

Vegetation consumed during the breeding season consistently includes grasses, sedges, leaves, and berries. Corn and winter wheat are commonly consumed during migration. In many regions, the winter diet is exclusively waste grains (for example, corn) and winter wheat. White-cheeked geese wintering in urban areas consume domestic grasses and also make field feeding flights to forage on waste grains and winter wheat. Although they primarily forage while walking, they frequently tip up for submersed aquatic vegetation.

RANGE AND HABITATS

Breeding: Cackling Geese breed in tundra regions of Alaska and western Canada. Canada Geese breed from tundra regions south into the contiguous United States. Many breed in the northern prairies and western mountain regions, and they have been introduced to urban and rural areas throughout most of the United States. Among Canada Geese, the Lesser subspecies is the most common tundra nester, and their breeding range overlaps with that of Cackling Geese. White-cheeked geese historically nested in eastern Russia and have been introduced to Europe and New Zealand. They are supreme generalists in their habitat selection. Breeding pairs use braided rivers, tundra, open boreal forests, temperate forests, prairies, coastal marshes, montane wetlands, golf courses, and urban parks. They are primarily upland nesters, but they also use muskrat mounds, hay bales, artificial nest structures (elevated or floating platforms), cliffs, and large stick nests (for example, old heron nests).
Migration: Migrants use a variety of habitats, including arctic river deltas, prairie rivers, wet meadows, prairie wetlands, playa wetlands, reservoirs, stock ponds, and agricultural fields.
Winter: Like other geese, they have historically been managed based on their wintering ground affiliation. Most wintering "populations" consist of multiple subspecies, and some include both Canada and Cackling Geese. Wintering populations tend to overlap greatly and shift in response to wetland conditions and food availability. White-cheeked geese winter along the west coast of Canada, sporadically across southern Canada, throughout the United States, and in Mexico. They use postharvest grain fields, rice fields, wet meadows, playa wetlands, rainwater basins, prairie rivers, shallow open wetlands, stock ponds, golf courses, municipal wetlands, urban

ponds, river deltas, freshwater impoundments, coastal estuaries, marshes, and bays. In portions of the Coastal Prairies they frequently associate with disked sorghum fields.

REPRODUCTION

Pair Bonds: White-cheeked geese typically pair in their second or third year. They tend to mate assortatively by size (by subspecies). Pair bonds are maintained for life, but they will form new pair bonds if their mate is lost. Fidelity to breeding areas is high.

Nesting: Females construct nests and incubate eggs. Females line nest bowls with down and vegetation, which they obtain from the vicinity of the nest site. Females lay one egg every 30–40 hours. Clutch size varies by subspecies, ranging from two to eight eggs. Males remain near nests during egg laying and incubation. During incubation breaks, males may guard nests or join females. Their incubation period is 25–28 days. Renesting is doubtful for tundra breeders, but white-cheeked geese breeding at lower latitudes may nest again if their initial attempt fails. They occasionally parasitize nests of other white-cheeked geese, and their eggs have been found in Osprey nests.

Goslings: Parents lead goslings permanently away from the nest within 48 hours after they hatch. Females brood goslings occasionally. Both parents defend goslings from potential predators. Goslings are capable of flight at 42–63 days; larger subspecies take longer to fledge than smaller subspecies. They migrate south with their parents.

APPEARANCE

Adults (*year-round*): Both Cackling and Canada Geese have similar plumages. They have a black neck (which has the appearance of a black sock pulled down to the base of the neck) with a white chin strap. Neck sock length varies among subspecies. The chin strap is broad and starts behind the eye, wrapping downward and under the chin, and in some individuals the white markings do not quite touch under the chin. The rump is large and white. They have black tails. In flight, a white V-shaped band is visible above the tail. Some individuals also have a partial white neck stripe separating their neck and breast. Breast, back, sides, and belly color in white-cheeked geese ranges from medium gray to dark brown. Feathers on their sides and back almost always have light edging. Body color of smaller subspecies tends to be darker than that of larger subspecies, although variation within subspecies is common. Occasionally the neck may be flecked with white throughout, and rarely the chin stripe may be dusky. All have black bills and feet.

Larger subspecies of Canada Geese are readily distinguishable from Cackling Geese by their size. Lesser Canada Geese, however, are only slightly different from Cackling Geese. Relative bill size is a useful means of separating Lesser Canada Geese from Cackling Geese, as bill size of Lessers tends to be large relative to their head size. However, like many traits, bill size is not diagnostic, as there is some overlap in sizes.

The smallest Cackling Goose, Ridgway's, averages 3.4 pounds (males), and the largest, Taverner's, averages 5.8 pounds (males). The smallest Canada Goose, Dusky, averages 7.1 pounds (males), and the largest, Giant, averages 10.7 pounds (males). There is overlap in the weights of many subspecies, particularly among males and females.

First winter: This plumage is largely identical to that of adults, except some first-winter geese may have a neck that is slightly brownish. They also have less contrast between the base of their neck and their body. That is, the line where their dark neck and grayish-brown body meet appears blurred instead of sharp.

SOURCES

INTRODUCTION: Scribner et al. 2003; Banks et al. 2004; Mlodinow et al. 2008. TEXAS DISTRIBUTION: Grieb 1970; Pulich 1988; Ballard and Tacha 1995; Ray and Miller 1997; Rusch et al. 1998; Esslinger and Wilson 2001; Lockwood and Freeman 2004; Baar et al. 2008; USFWS 2008. TEXAS HARVEST: Kruse 2007. LONGEVITY: Lutmerding and Love 2011. POPULATION STATUS: Rutherford 1965; Grieb 1970; Moser et al. 2004; NAWMP 2004; USFWS 2011. DIET: Eggeman et al. 1989; Ray and Higgins 1993; Reed et al. 1996; Rusch et al. 1998; Gates et al. 2001; Moser et al. 2004. RANGE AND HABITATS: Hanson and Eberhardt 1971; Bellrose 1980; Petersen 1990; Ballard and Tacha 1995; Carrière et al. 1999; Dickson 2000; Mowbray et al. 2002; Eichholz and Sedinger 2006. REPRODUCTION: Brakhage 1965; Hanson 1965; Raveling 1969, 1988; Mickelson 1975; Cooper 1978; Bellrose 1980; Zicus 1984; Batt et al. 1992; Rusch et al. 1998; Carrière et al. 1999; Dickson 2000; Mowbray et al. 2002. APPEARANCE: Bellrose 1980; Mowbray et al. 2002; Hanson 2006; Mlodinow et al. 2008.

BRANT

Branta bernicla

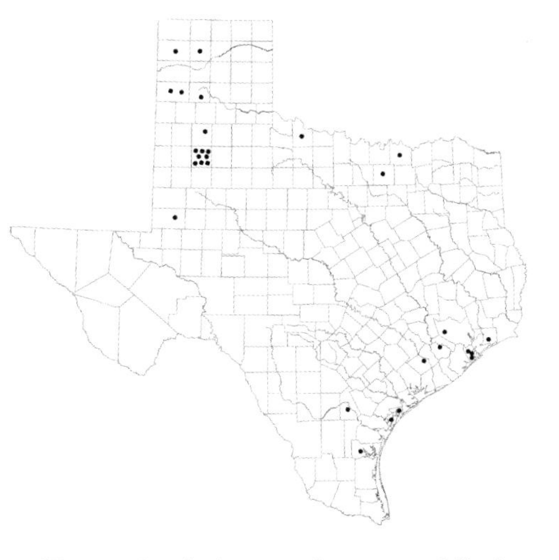

Brant are rare to casual visitors to Texas. There have been 29 well-documented records, and these have consisted of Light-bellied Brant (*B. b. hrota*) and Black Brant (*B. b. nigricans*). Most records are from the High Plains and Coastal Prairies. The consensus is that Brant occur in Texas much more regularly but go undetected in large flocks of white-cheeked geese, particularly in northwestern Texas. There are many secondhand reports of Brant taken by hunters that are not fully documented as records. One banded Brant has been recovered in Texas; it was banded on Baffin Island (Nunavut) in 1993 and recovered near the Texas coast approximately five years later.

Brant breed along the arctic fringe of North America, Europe, and Asia. In Europe they are known as Brent Geese. Brant are typically divided into three different subspecies. Two of these, Black Brant and Light-bellied Brant (also called Atlantic Brant), occur in North America. Brant breed from Baffin Island west into Alaska

Light-bellied Brant (also called Atlantic Brant). *Photograph by Mark W. Lockwood, February 8, 2006, near Plymouth, Mass.*

Black Brant (immature). *Photograph by Mark W. Lockwood, November 10, 2007, Andrews, Andrews County, Texas.*

and also in northern Greenland. They nest on the ground, either dispersed or in colonies. Nests are found in marshes and grassy flats above the tide line and on gravel bars. Nests are typically exposed. During winter, Black Brant are found in Izembek Lagoon in western Alaska and scattered along the Pacific coast from southeastern British Columbia to Baja California. One of the largest concentrations winters off the coast of Sonora. On the Atlantic coast, wintering Light-bellied Brant are scattered from Massachusetts to North Carolina. They are primarily herbivores, foraging on a wide array of saltmarsh plants during the breeding season and on eelgrass, widgeongrass, sea lettuce, cultivated grasses, and tame clover during winter.

Like most geese, male and female Brant are identical in appearance. Adults have a black head, neck, breast, and upper back. Their upper neck has a broad but incomplete streaky white band. They have a brownish back and a white rump. Light-bellied Brant have a light brown belly that becomes whiter near the vent. Black Brant have blackish bellies. First-winter Brant do not have white markings on their neck. Average fall weights of males and females are 4.0 and 3.6 pounds, respectively.

From 2000 to 2003 the annual Brant estimate was 286,520, including all subspecies. In the past Brant suffered large population declines that were associated with severe declines in eelgrass on their wintering grounds. Their population is currently stable.

SOURCES

Cottam et al. 1944; Bellrose 1980; Smith et al. 1985; Baldwin and Lovvorn 1994; Reed et al. 1998; Buckley et al. 2004; Lockwood and Freeman 2004; NAWMP 2004; Ward et al. 2005; USGS 2010.

TRUMPETER SWAN

Cygnus buccinator

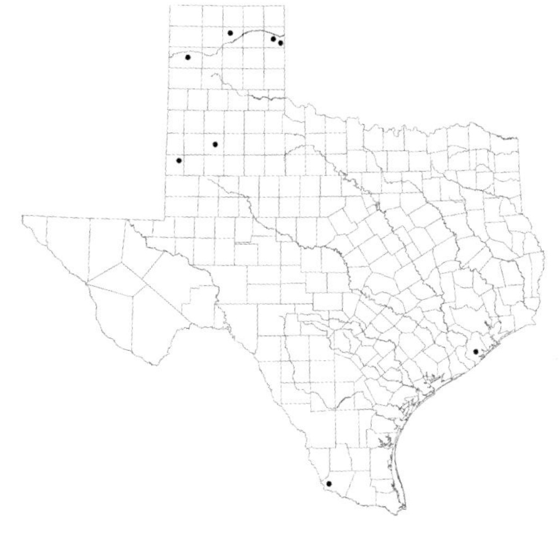

Historically, Trumpeter Swans were a common winter visitor to Texas and may have been abundant near the coast. They have been very rare visitors since the late 1800s. Virtually all records during the 1990s involved neck-collared birds that originated from reintroduction efforts in midwestern states. Unmarked Trumpeter Swans in Texas may also originate from these reestablished populations, as neck-collared adults that were part of a reintroduction program in Iowa have been observed in Texas with unmarked young. Furthermore, reintroduction programs in Minnesota and Wisconsin stopped mark-

Trumpeter Swan (adult). *Photograph courtesy of Glenn Bartley/VIREO, taken November 10, 2007, Victoria, British Columbia, Canada.*

ing nestlings by 2000 because their populations had become self-sustaining. It is also plausible that some Trumpeter Swans in the western Panhandle may originate from the northern Rocky Mountains, as is evidenced by a cygnet (young swan) banded in Wyoming that was later found in Oldham County after it collided with a wire. Trumpeter Swans may occur throughout the state.

Once abundant, Trumpeter Swans were severely reduced by habitat degradation and market hunting. From 2001 to 2003 their population averaged 23,647 annually.

Their numbers hit a modern high in 2005 with 34,803 swans. Their population trend is increasing.

Trumpeter Swans breed in Alaska and western Canada and locally in the mountain West, northern plains, and midwestern United States. Breeding pairs use freshwater marshes, prairie wetlands, lakes, ponds, and rivers. They nest on the ground, in emergent vegetation (overwater nests), or in moist areas along wetland edges. They may locate nests on beaver lodges or muskrat mounds. Females typically lay 4–6 eggs. Both members of the pair share incubation duties. They mate for life but will re-pair if their mate is lost. Breeding Trumpeter Swans primarily forage on submersed aquatic vegetation (for example, pondweed) and other wetland plants.

Most winter in western Canada and in the northwestern United States. Scattered flocks also winter in the Intermountain West, and small numbers winter in the central United States. Trumpeter Swans that winter in the central United States likely originate from northern plains or midwestern states. Wintering swans use shallow estuaries, freshwater ponds, lakes, and rivers. Their diet includes aquatic vegetation, terrestrial grasses, and agricultural grains.

Adults are white and have a black eye, black bill, and black legs and feet. Adults may have a few gray feathers on their head and neck. Males and females are identical. First-winter Trumpeter Swans have a gray-tinged head, neck, and back. Trumpeter Swans are the largest North American waterfowl and one of the heaviest flying birds in the world, weighing up to 30 pounds.

SOURCES

Hansen et al. 1971; Bellrose 1980; Mitchell 1994; Matteson et al. 1995; Burgess and Burgess 1997; Seyffert 2001; Lockwood and Freeman 2004; NAWMP 2004; Dean 2005; Moser 2006.

TUNDRA SWAN

Cygnus columbianus

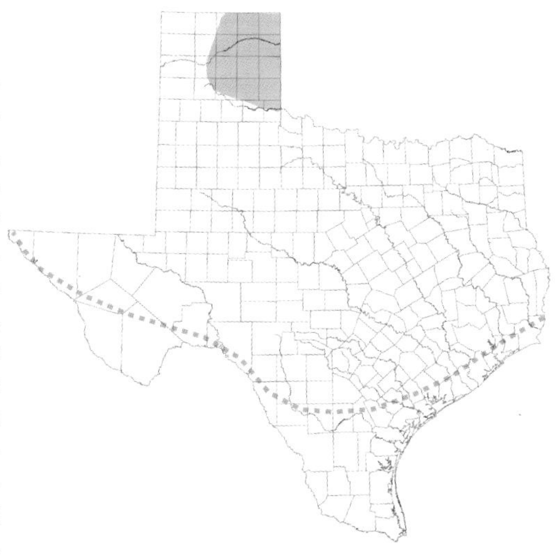

Tundra Swans are rare to very rare in Texas. They potentially occur throughout the state, but are most often observed in the northeastern High Plains and northern Rolling Plains. Most Tundra Swans found in Texas are first-winter birds, and flocks of more than three are rare. Larger groups have been reported irregularly in Hemphill and Donley Counties. Most observations occur between late October and mid-March. They were poten-

Tundra Swan (adult). *Photograph courtesy of Arthur Morris/VIREO, taken June 15, 2003, Nome, Alaska.*

tially more common in Texas before 1900. Tundra Swans were collected as part of ornithological expeditions in Lee County in 1888 and near Brownsville in 1878. Multiple sources also suggest they were common along the coast of nearby Tamaulipas, Mexico.

Tundra Swans breed in arctic regions of the United States and Canada. Breeding pairs use tundra wetlands, particularly those with abundant pondweed. They mate for life but will re-pair if their mate is lost. They nest on the ground, often on hummocks in wet meadows and on islands. Females typically lay 3–5 eggs. Both males and females incubate.

Most Tundra Swans migrate through interior North America in family groups. They winter along the west coast of the United States, locally in the Intermountain West, and along the central Atlantic coast (Chesapeake Bay area). They were historically abundant in Mexico, although only a few currently winter there. Wintering birds mainly use shallow estuaries, freshwater ponds, lakes, and rivers. Their diet includes leafy vegetation, tubers, rhizomes, submersed aquatic vegetation (for example, pondweed), agricultural grains, and mollusks.

Males and females average 15.9 and 13.9 pounds, respectively, and are identical in appearance. They are white and have a black eye, black bill, and black legs and feet. Most Tundra Swans have a small yellow marking in front of their eye. First-winter Tundra Swans have a gray-tinged head, neck, and back. They generally have whiter body plumage than first-winter Trumpeter Swans. In 2011 their population was 174,000. Since 2000 their numbers have fluctuated from year to year, but their population is stable.

This species was formerly known as Whistling Swan. The name was changed when that taxon was lumped with Bewick's Swan (*C. c. bewickii*), which breed in Europe and Asia.

SOURCES

Merrill 1878; Singley 1892; Limpert et al. 1987; Pulich 1988; Limpert and Earnst 1994; Drewien and Benning 1997; Seyffert 2001; Earnst and Rothe 2004; Lockwood and Freeman 2004; NAWMP 2004; Badzinski 2005; USFWS 2011.

MUSCOVY DUCK

Cairina moschata

Domestic Muscovy Ducks are common throughout the Americas and much of the world. Today, domestic forms are larger than their wild counterparts, with bright red facial skin and more white plumage. The exact timing of their domestication is elusive. Anthropological evidence suggests that Muscovy Ducks may have been domesticated and even traded in western Ecuador over 1,100 years ago. Specimens have also been found at human burial sites in Ecuador that are over 2,600 years old.

TEXAS DISTRIBUTION

Texas is the only state that is part of the natural range of wild Muscovy Ducks. They have been reported along the Rio Grande in Maverick, Zapata, Starr, and Hidalgo Counties, but they are most common in Starr and Hidalgo. Nesting was first confirmed in Texas in Hidalgo County in 1994.

TEXAS HARVEST

There is no evidence that wild Muscovy Ducks are harvested in Texas.

POPULATION STATUS

There is no population survey or population estimate for Muscovy Ducks.

Muscovy Ducks (male and female). *Photograph courtesy of Patricio Robles Gil/VIREO, taken in Tamaulipas, Mexico.*

DIET

Their diet is varied. In Mexico their summer diet includeds corn, soldier flies, waterlily seeds, and mangrove seeds. Muscovy Ducks are also reported to forage on small crabs, termites, and roots.

RANGE AND HABITATS

Breeding and Wintering: Muscovy Ducks are found from the Rio Grande to Argentina. They are nonmigratory. Breeding birds use sluggish rivers, streams, swamps, freshwater wetlands, and ponds located near woodlands. They nest in large tree cavities and in artificial nest structures (nest boxes). They occasionally nest on the ground, in palm fronds, and in crevices within caves. Nonbreeding Muscovy Ducks use the same types of wetlands as do breeding birds but are also found in coastal swamps, marshes, and lagoons.

REPRODUCTION

Pair Bonds: Weak male-female associations may develop for brief periods during the breeding season, but Muscovy Ducks are largely promiscuous.
Nesting and Ducklings: In Tamaulipas, Muscovy Ducks nest from April through July,

but their nesting season may coincide with the rainy season in more tropical areas. They add down to their nests but do not carry vegetation to the nest cavity. Eggs are laid at a rate of 1 per day, and only females incubate. Clutch size varies from 9 to 15 eggs. Their incubation period is about 30 days. Renesting has been documented. Duckling ecology is unstudied.

APPEARANCE

Muscovy Ducks are brownish black to black with conspicuous white patches on their upper wing. Their body, neck, and head often have purple or green iridescence. Males have bare skin around the base of their bill and their eyes, and their facial skin is black with some reddish areas, usually between their eyes and bill, and typically has fleshy caruncles. Females do not have bare skin around their eyes, and their caruncles are reduced or absent. There are considerable size differences between female and male Muscovy Ducks. Females weigh 2.4 to 3.2 pounds and males weigh 4.4 to 8.8 pounds.

SOURCES

INTRODUCTION: Hesse 1980; Stahl et al. 2006. TEXAS DISTRIBUTION: Brush and Eitniear 2002; Lockwood and Freeman 2004. DIET: Johnsgard 1975, 1978; Woodyard and Bolen 1984. RANGE AND HABITATS: Lack 1974; Johnsgard 1975, 1978; Woodyard and Bolen 1984; Markum and Baldassarre 1989; Eitniear et al. 1998. REPRODUCTION: Bolen 1971b; Johnsgard 1975, 1978; Woodyard and Bolen 1984; Markum and Baldassarre 1989. APPEARANCE: Leopold 1959; Johnsgard 1975, 1978

WOOD DUCK

Aix sponsa

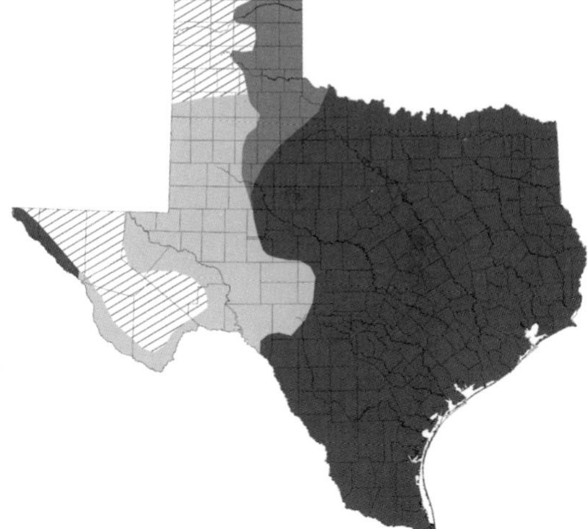

Wood Ducks are the only species of North American waterfowl that double broods frequently or produces two successful broods per season. Double brooding has been recorded throughout the lower two-thirds of their range, including Texas. Depending on location and year, 0–18 percent of females may double brood.

TEXAS DISTRIBUTION

Breeding: Wood Ducks commonly nest throughout the eastern three-fourths of Texas, but nesting densities are highest in the Pineywoods and Post Oak Savannah–Blackland Prairies. Localized breeding may take place in the remainder of the state if adequate wetlands and nest cavities are available. Wood Ducks are uncommon residents along the Rio Grande in El Paso County.

Wood Duck (male and female; note the feathers on the male's head are windblown in an atypical way). *Photograph by Raymond S. Matlack, January 6, 2007, Amarillo, Potter County, Texas*

Migration and Winter: Texas hosts both resident and migrant Wood Ducks, so it is difficult to gauge the progress of migration. They may occur statewide during fall and winter but are typically only common in the eastern portion of the state. Depending on the calculation method used, density estimates in flooded forests in Nacogdoches County ranged from 0.45 to 1 duck per 2.47 acres. Wood Ducks are likely underestimated in the Texas Mid-winter Waterfowl Survey, as forested wetlands are not conducive to aerial surveys.

TEXAS HARVEST

From 1999 to 2006, Wood Duck harvest in Texas averaged 64,180 annually. This was approximately 5 percent of their annual US harvest.

LONGEVITY

The average life span of adult males and females is 1.7 and 1.4 years, respectively. The longevity record for a wild Wood Duck is 22 years, six months.

POPULATION STATUS

Wood Ducks have long rebounded from once dreadfully low population levels. Estimates based on harvest and banding data suggest the continental population may exceed 2.9 million. They have a healthy, if not increasing, population. Their populations are monitored through banding efforts, regional surveys, regional nest box monitoring programs, and harvest surveys.

DIET

Wood Ducks are omnivorous. In Missouri, invertebrates accounted for 76 percent of the diet of laying females; important invertebrates were scavenger beetles, diving beetle larvae, crane fly larvae, horse fly larvae, and midges. Crayfish and snails are also consumed by breeding Wood Ducks. In Cherokee County, Texas, their fall and winter diet was 97 percent plant material, with acorns accounting for 78 percent of their diet. Wintering Wood Ducks may also consume corn and soybeans.

RANGE AND HABITATS

Breeding: Wood Ducks breed in southeastern Canada, the eastern United States, and Cuba. Nesting also occurs in southwestern Canada, the Pacific Northwest, and California and in forested riparian zones of central North America. Isolated breeding populations occur in urban areas outside of their traditional range and in Mexico. Breeding pairs are found on rivers, streams, riparian wetlands, beaver ponds, shrub-scrub wetlands, ponds, lakes, temporary wetlands, marshes, moist-soil management wetlands, and forested wetlands. Preformed, existing cavities are required for nesting, such as those excavated by Pileated Woodpeckers. Nest cavity entrances range from 2 feet to more than 55 feet above ground. Wood Ducks readily use nest boxes. They occasionally choose chimneys as nest sites. Although extremely rare, nesting in hay lofts, on the ground, and in large stick nests (for example, old raptor nests) has been documented. Nest cavities may be over land or water.
Migration and Winter: The breeding, migration, and wintering ranges of Wood Ducks greatly overlap. Migrants regularly use rivers, riparian wetlands, and beaver ponds. Fall migrants commonly forage in agricultural fields. Their wintering range extends into South Texas, the southwestern United States, and Mexico. Wintering Wood Ducks have a strong affinity for bottomland hardwoods and forested wetlands; they also use beaver ponds, freshwater marshes, reservoirs, sheet water, flooded and dry agricultural fields, and moist-soil management wetlands. They commonly use shrub-scrub wetlands as communal roost sites.

REPRODUCTION

Pair Bonds: Wood Ducks are seasonally monogamous. In Missouri, pair bond formation began in September, and 95 percent of females were paired by May. Females select new mates each year. Males follow their mates to breeding locations. Females often use the same nest cavity from one year to the next. Pair bonds dissolve during incubation.
Nesting: Because cavities are often scarce, reports of Wood Ducks usurping or attempting to usurp the nests of other cavity-nesting species (for example, Pileated

Woodpeckers) are common. There are also reports of Wood Ducks losing their nests to other cavity-nesting species. They do not carry vegetation to the nest cavity, but they do add down. Eggs are laid at a rate of 1 per day, and the clutch size is about 10–12 eggs. Incubation begins before the clutch is complete. Only females incubate, taking incubation breaks in the morning and evening. The incubation period lasts about 31 days. In eastern Texas, nest success in artificial cavities ranged from 64 percent to 82 percent during a two-year study. Renesting is common if nests are lost. Females may also renest if their ducklings are lost, and double brooding is regular.

Wood Ducks are renowned for egg dumping, or parasitizing the nests of other Wood Ducks. In nest boxes parasitism rates often exceed 40 percent. Nests containing large numbers of eggs are commonly referred to as dump nests. Four hens have been reported to lay in a single nest in one day. In extreme cases, parasitized nests may contain more than 40 eggs. Wood Ducks have also been reported to parasitize nests of other species, including cavity-nesting owls.

Ducklings: Females lead ducklings permanently away from the nest within 24 hours after they hatch. Females fly from the nest and give vocal cues to encourage ducklings to exit. The fall from the nest does not injure the ducklings. Only females care for the young; they brood ducklings, feign injury to distract predators away from ducklings, and frequently lead ducklings to new or different wetlands. Females may lead ducklings overland or through a series of connected wetlands. In Brazos and Burleson Counties, distances between the wetland initially used by ducklings and the wetland in which the brood was eventually reared ranged from less than 200 feet to over 7 miles. Ducklings fledge at 56–70 days. Females often abandon their ducklings before they fledge.

APPEARANCE

Breeding: Wood Ducks exhibit strong sexual dimorphism. Adult males have an iridescent green head and a large iridescent green and purple crest. A thin white line starts at the bill and arches upward and back over the top of the head and then down the top of the crest; a second white line starts behind the eye and arches backward along the bottom of the crest. The males have a white chin patch. A narrow white arc extends upward from the chin patch toward the eye, and a second arc forms a partial neck ring. Their breast is chestnut brown with rows of white spots. Their sides are yellow to gold. Their breast and sides are demarcated by two adjacent vertical stripes, one white and one black; the white stripe is to the front. Their back, rump, and tail are iridescent and dark, and the area under their tail is reddish. Their belly is white. They have a brightly colored yellow, red, white, and black bill. Adult males reach their peak weight of about 1.7 pounds during late winter.

The neck, head, and crest of adult females are gray with a mild purplish sheen. Their crest is slight compared to that of males. Around their eye females have an oval white area, which tapers to the back. Their breast is grayish brown with tan streaks, and their sides are grayish brown. Their back is gray to olive with a slight sheen. Females have a gray bill. Adult females reach their peak weight of about 1.6 pounds during egg laying.

Nonbreeding: In males, this plumage is like that of females. They take on a grayer appearance and lose their long crest but retain their white throat patch. In females, this plumage is similar to their breeding plumage, but the sheen on their back and head is not as glossy.

SOURCES

INTRODUCTION: Fredrickson and Hansen 1983; Bellrose and Holm 1994. TEXAS DISTRIBUTION: Bellrose and Holm 1994; Lockwood and Freeman 2004; Ransom and Frentress 2007; Whiting and Cornes 2009. TEXAS HARVEST: Kruse 2007. LONGEVITY: Bellrose and Holm 1994; Lutmerding and Love 2011. POPULATION STATUS: Bellrose and Holm 1994; Hepp and Bellrose 1995. DIET: Landers et al. 1976; Drobney and Fredrickson 1979; Allen 1980; Delnicki and Reinecke 1986; Bellrose and Holm 1994; Hepp and Bellrose 1995; Havara 1999a; Miller et al. 2003. RANGE AND HABITATS: Steward 1971; Soulliere 1980; Frentress 1987; Thompson and Baldassarre 1988; Bellrose and Holm 1994; Hepp and Bellrose 1995; Perezgasga 1999; Hartke and Hepp 2004. REPRODUCTION: Leopold 1951; Bellrose et al. 1964; Bolen and Cain 1968; Morse and Wight 1969; Drobney 1980; Armbruster 1982; Haramis and Thompson 1985; Semel and Sherman 1986; Hepp et al. 1987; Kennamer and Hepp 1987; May and Kroll 1989; Ridlehuber et al. 1990; Hepp and Kennamer 1992; Bellrose and Holm 1994; Dugger et al. 1994; Hepp and Bellrose 1995; Manlove and Hepp 2000; Conner et al. 2001. APPEARANCE: Drobney 1982; Bellrose and Holm 1994; Hepp and Bellrose 1995; Hipes and Hepp 1995; Havara 1999a.

GADWALL

Anas strepera

Gadwalls have the greatest salt tolerance of any North American dabbling duck. Northern Pintails often winter in hypersaline (> 35 ppt) environments, such as the Laguna Madre, but breed in freshwater wetlands. Young Mottled Ducks can tolerate salinities up to 11 ppt, but their survival decreases at greater salinities. However, Gadwalls often breed around areas with high salinities. At California's Mono Lake (87 ppt), a long-term study found high brood survival. Salt tolerance in Gadwall ducklings increased as they aged. Newly hatched ducklings were restricted to freshwater areas that occurred along the lake's edge, but within one week they were able to abandon those sites for more saline areas.

TEXAS DISTRIBUTION

Breeding: There are approximately 15 breeding records for Gadwalls in Texas. Most of these involve duckling sightings in the High Plains. There are also 2 potential breeding records from Dallas County and one from Bexar County. Although uncommon,

Gadwall (male). *Photograph by Greg Lasley, December 20, 2007, Austin, Travis County, Texas.*

Gadwall (female). *Photograph by Trey Barron, December 25, 2009, Amarillo, Potter County, Texas.*

male-female pairs occur annually in the High Plains in both May and June. Gadwalls are irregularly observed throughout the rest of the state during summer.
Migration: Gadwalls migrate through all Texas counties. In the High Plains their numbers start building in early October and peak in early November. Most are gone from the High Plains by December. In the Rolling Plains, Post Oak Savannah–Blackland Prairies, and Pineywoods the first migrants arrive in September; they are most common in the northern Rolling Plains and Post Oak Savannah–Blackland Prairies between mid-October and April. They begin arriving in the Coastal Prairies and Coastal Sand Plain in October; numbers in these areas jump greatly in November, remain high through March, and decline in April. In the Post Oak Savannah–Blackland Prairies and Pineywoods their numbers begin declining in late April or May. In the High Plains, spring migrants appear in early March, peak in early April, and decline during May. Gadwall abundance in the High Plains is greater in spring than in fall.
Winter: From 2000 to 2008, Gadwalls averaged 728,422 during the Texas Mid-winter Waterfowl Survey. They winter throughout the state but are most common in the Coastal Prairies, Post Oak Savannah–Blackland Prairies, South Texas Brush Country, and Rolling Plains (TPWD unpublished).

TEXAS HARVEST

In most years, more Gadwalls are harvested by Texas hunters than any other duck. From 1999 to 2006, Texas harvest averaged 247,214 annually. This accounted for about 16 percent of their annual US harvest.

POPULATION STATUS

In 2011 the estimated abundance of Gadwalls was 3.3 million. From 1955 to 2011, populations fluctuated from a low of 502,000 to a high of 3.9 million. The North American Waterfowl Management Plan's population goal is 1.5 million. Gadwalls have been above this level since 1990.

DIET

Gadwalls forage heavily on algae at all times of the year. Algae, invertebrates, and submersed aquatic plants accounted for most of their diet during the breeding season. Gadwalls wintering in coastal Louisiana commonly consumed algae and submersed aquatic plants. On the Texas coast, widgeongrass is likely an important food. Seeds are consumed in most studies, but rarely are they a dominant food item. Likewise, foods from agricultural crops are infrequently reported in Gadwalls' diet.

RANGE AND HABITATS

Breeding: Gadwalls breed in North America, northern Europe, and northwestern Asia. In North America the highest breeding densities occur in the Prairie Pothole Region. During the last century, their breeding range expanded westward into the Pacific Northwest and eastward into New England and Canada's eastern provinces. Pairs settle in on moderately to heavily vegetated seasonal and semipermanent wetlands. In areas where natural wetlands are scarce, stock ponds may be used. They are upland nesters. They readily nest in heavy grassland cover, dry seasonal wetlands, roadsides, islands, and odd areas (for example, overgrown rock piles and fencerows). Nests are occasionally found in short brush cover. They tend to avoid nesting in croplands.

Migration and Winter: Migrants occur throughout much of the United States. Gadwalls winter along the East and West Coasts, throughout the South, and locally in the West. They also winter throughout Mexico. They use shallow to moderately deep wetlands, particularly those dominated by submersed aquatic and floating plants. Wetlands used include reservoirs, stock ponds, beaver ponds, and fresh, intermediate, and brackish coastal marshes. In Texas Gadwalls are the most common duck occurring on flood prevention lakes in the Post Oak Savannah–Blackland Prairies.

REPRODUCTION

Pair Bonds: Pair bond formation begins during fall migration. In coastal Louisiana over 80 percent of Gadwall females were paired in November. Females choose their mates. Re-pairing between males and females in successive years is extremely rare. Once paired, males follow their mates to breeding areas. Females have a high degree of fidelity to their breeding grounds. Pair bonds dissolve during incubation.

Nesting: Only female Gadwalls construct nests and incubate eggs. Females line nest bowls with down and litter from herbaceous vegetation, which is obtained at the nest site. Eggs are laid at a rate of 1 per day. Clutch size varies from 9 to 11 eggs. The incubation period lasts about 26 days. If the initial clutch is lost, renesting is common. Up to four nesting attempts may be made in a breeding season. Females do not attempt to renest after a clutch has hatched, even if their ducklings are lost. When nesting in high densities, females regularly parasitize nests of other ducks.

Ducklings: Females brood ducklings, lead them to food, and lead them away from predators. Ducklings have been known to travel over one mile between hatching and fledging. Young fledge at about 50 days.

APPEARANCE

Breeding: At a distance, adult male Gadwalls have an overall appearance that is gray except for a black rump. They have a tan head and neck, and their head plumage often appears puffed out. Their breast is black with white and buff markings. Their sides and back appear gray; their back has long feathers, which are dark gray with buff edging. Their belly is white. Adult females have a tan head and neck and a dark eye-line. Their body feathers are dark brown with buff edging, which gives them a mottled brown appearance. Their belly is white.

Males have a slaty blue to black bill, which has yellow to orange markings on its sides and underparts. Females have a dark brown bill with yellow to orange sides; bills of females often have small dark splotches. Adult males and females average 2.1 and 1.8 pounds, respectively.

Nonbreeding: In late summer and fall both males and females are mottled brown.

SOURCES

INTRODUCTION: Moorman et al. 1991; Austin and Miller 1995; Jehl 2005. TEXAS DISTRIBUTION: Hobaugh and Teer 1981; Pulich 1988; Seyffert 2001; White 2002; Ray et al. 2003; Lockwood and Freeman 2004; Baar et al. 2008; USFWS 2008. TEXAS HARVEST: Kruse 2007. POPULATION STATUS: NAWMP 2004; USFWS 2011. DIET: Landers et al. 1976; Serie and Swanson 1976; Paulus 1982; LeSchack et al. 1997. RANGE AND HABITATS: Higgens 1977; Kantrud and Stewart 1977; Ruwaldt et al. 1979; Bellrose 1980; Paulus 1983; Hines and Mitchell 1983; Kaminski and Prince 1984; Kantrud et al. 1989; Lokemoen et al. 1990; Greenwood et al. 1995; LeSchack et al. 1997. REPRODUCTION: Rienecker and Anderson 1960; Keith 1961; Gates 1962; Oring 1969; Blohm 1979; Bellrose 1980; Hines and Mitchell 1983; Paulus 1983; Fedynich and Godfrey 1989; Lokemoen et al. 1990; Lokemoen 1991; Greenwood et al. 1995; LeSchack et al. 1997; Sayler and Willms 1997; Pietz et al. 2003. APPEARANCE: Bellrose 1980; LeSchack et al. 1997; Havera 1999a.

EURASIAN WIGEON

Anas penelope

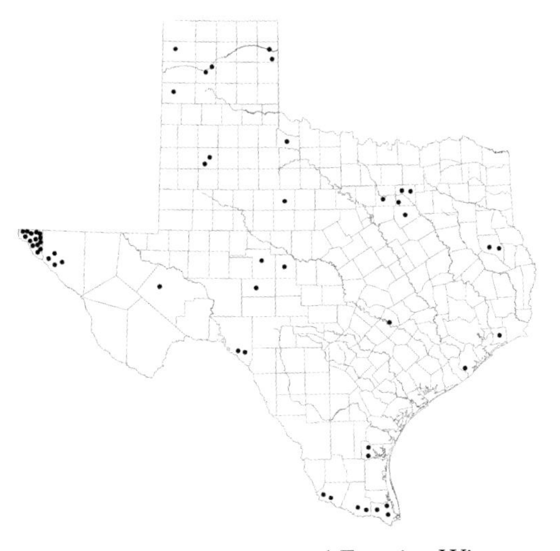

Eurasian Wigeon are casual visitors to Texas. As of spring 2012 there have been 53 well-documented records. These are from all areas of the state, but the majority are from El Paso and Hudspeth Counties. Some individuals have overwintered in Texas. Dates of occurrence range from early November to late April. All Texas records have involved males. Eurasian Wigeon are typically associated with flocks of American Wigeon. There have also been several Eurasian Wigeon x American Wigeon hybrids documented in the state.

Eurasian Wigeon breed across northern Europe and Asia in boreal forests and in subarctic and arctic regions. They are highly herbivorous, often grazing on wheat. In breeding plumage males are distinguishable from American Wigeon by their cinnamon head and neck and their buff forehead. In flight, on water, and in hand, female Eurasian Wigeon are similar in appearance to female American Wigeon. Some authorities describe female Eurasian Wigeon as having a browner head and neck than those of female American Wigeon, but it is doubtful these subtleties could be distinguished in the field.

SOURCES

Johnsgard 1978; Gooders and Boyer 1986; Lane et al. 1998; Lockwood and Freeman 2004.

Eurasian Wigeon (male). *Photograph courtesy of G. Bartley/VIREO, taken January 17, 2007, Victoria, British Columbia, Canada.*

AMERICAN WIGEON

Anas americana

American Wigeon have the most herbivorous diet of the dabbling ducks. They specialize on submersed aquatic vegetation, soft parts of emergent vegetation, and leafy herbaceous plants. They forage in both wetlands and uplands. Surprisingly, there are few reports of American Wigeon depredating winter wheat. In contrast, their Old World counterparts, Eurasian Wigeon, regularly graze on wheat and are often linked to crop damage in Asia.

TEXAS DISTRIBUTION

Breeding: Pairs are occasionally observed in early summer in the High Plains and in El Paso County; however, there is no evidence that they breed in Texas.

Migration: American Wigeon migrate through all Texas counties. In the High Plains migrants arrive in early August, and their numbers peak in early November. Their abundance declines sharply during late November, although many remain in the High Plains throughout winter. In the Rolling Plains, Post Oak Savannah–Blackland Prairies, and Pineywoods the first migrants arrive in August, and they are common by September or October. Many are present in the Coastal Prairies and Coastal Sand Plain during September; numbers in these two regions peak in November or December and then gradually decline through spring. In the High Plains, spring migrants begin returning in early February, and their numbers peak in late March. Most are gone from northern portions of the state by late May.

Winter: From 2000 to 2008 American Wigeon averaged 287,832 during the Texas Mid-winter Waterfowl Survey. They winter throughout the state but are most common in the Rolling Plains, Coastal Prairies, South Texas Brush Country, and Post Oak Savannah–Blackland Prairies (TPWD unpublished).

American Wigeon (male and female). *Photograph by Raymond S. Matlack, January 25, 2007, Canyon, Randall County, Texas.*

TEXAS HARVEST

From 1999 to 2006 American Wigeon harvest in Texas averaged 108,635 annually. This accounted for about 15 percent of their annual US harvest.

LONGEVITY

The longevity record for a wild American Wigeon is 21 years, four months.

POPULATION STATUS

In 2011 their estimated abundance was 2.1 million. From 1955 to 2011, populations fluctuated from a low of 1.7 million to a high of 3.8 million. American Wigeon have been below the North American Waterfowl Management Plan's population objective of 3 million since 1991.

DIET

Laying females in Saskatchewan consumed 59 percent vegetation and 41 percent animal matter. Migrants in Illinois foraged almost entirely (99 percent) on vegetation. Past studies suggested corn accounted for 88 percent of food consumed by migrating and wintering American Wigeon in the High Plains. However, this estimate was likely biased high because birds were collected upon their return from field feeding. In Texas coastal wetlands, widgeongrass is likely important. In Humboldt Bay, California, their diet was mostly eelgrass and clover.

RANGE AND HABITATS

Breeding: American Wigeon breed from northern Colorado northwest to Alaska and northeast to Ontario. They also breed in California and in Canada's Maritime Provinces. Breeding pairs use wet meadows, prairie rivers, large marshes, stock tanks, and seasonal, semipermanent, and permanent wetlands. They are upland nesters, using both herbaceous vegetation and brush for nest sites.
Migration and Winter: Migrants occur throughout much of North America. American Wigeon winter in southwestern Alaska, western Canada, much of the continental United States, Mexico, Central America, northern South America, and the Caribbean. They frequent wetlands that have abundant submersed aquatic vegetation. They use reservoirs, stock ponds, sluggish rivers, river deltas, playa wetlands, flooded agriculture fields, freshwater marshes, coastal marshes, estuaries, and bays. They are common in urban wetlands.

REPRODUCTION

Pair Bonds: Pair formation begins during fall migration. In the Texas High Plains, 27 percent of females were paired in November, 45 percent in January, and 81 percent in March. Females choose their mates. Once paired, males follow their mates to breeding areas. Adult females return to areas where they have previously bred at rates about 39–44 percent. Pair bonds are temporary, dissolving during incubation.
Nesting: Nest construction and incubation are solely activities of the female. Females line nest bowls with down and with plant litter, which is obtained from the immediate vicinity of the nest. Eggs are laid at a rate of one per day. The clutch size averages

eight to nine eggs. Incubating females spend approximately 88 percent of their time on the nest, and their incubation period is about 23–25 days. They are not known to parasitize other nests.
Ducklings: Only females care for the young. They lead ducklings permanently away from the nest within 24 hours after they hatch. Females brood ducklings, lead them to food, and often lead them overland to larger, more permanent wetlands. Young are capable of flight at 44–47 days.

APPEARANCE

Breeding: Adult males have a light gray head and a white forehead, which is where they get the colloquial name baldpate. A green, crescentlike line starts just forward of the eye and arches back and downward. They have a buffy neck that blends into a pinkish-rose breast. A patch of white plumage separates their pinkish sides and black rump. Males have a white belly. Their tail feathers are pale. Their back has both long tan feathers and long black feathers, the latter of which have light edging.

Adult females have a brown body and a light brown head and neck; their head, neck, and back are heavily speckled with black. Their breast and sides are pinkish tan and their belly is light. They have a dark rump and grayish-brown tail feathers. Both sexes have a small, pale blue bill with a black tip. Their bill is short, narrow, and gooselike. In the High Plains, adult males and females averaged 1.7 and 1.6 pounds, respectively.
Nonbreeding: The nonbreeding plumage of both males and females is similar to the breeding plumage of females.

SOURCES

INTRODUCTION: Moore 1980; Baldwin and Lovvorn 1992; Lane et al. 1998; Mowbray 1999; USDA 2009. TEXAS DISTRIBUTION: Pulich 1988; White 2002; Benson and Arnold 2003; Ray et al. 2003; Lockwood and Freeman 2004; Baar et al. 2008; USFWS 2008. TEXAS HARVEST: Kruse 2007. LONGEVITY: Lutmerding and Love 2011. POPULATION STATUS: NAWMP 2004; USFWS 2011. DIET: Yocum and Keller 1961; Moore 1980; Baldassarre and Bolen 1984; Smith et al. 1989; Havera 1999b; Mowbray 1999; Hartke et al. 2009. RANGE AND HABITATS: Lokemoen 1973; Bellrose 1980; Turnball and Baldassarre 1987; Thompson and Baldassarre 1991; Baldwin and Lovvorn 1992; Higgens et al. 1992; Ball et al. 1995; Sedinger 1997; Elphick and Oring 1998; Mowbray 1999; Sovada et al. 2001; Bogiatto and Karnegis 2006. REPRODUCTION: Evans et al. 1952; Soutiere et al. 1972; Bellrose 1980; Wishart 1983; Batt et al. 1992; Arnold and Clark 1996; Mowbray 1999. APPEARANCE: Bellrose 1980; Mowbray 1999; Havera 1999a; Rhodes et al. 2006.

AMERICAN BLACK DUCK

Anas rubripes

American Black Ducks are rare winter visitors to Texas. Recent sightings have largely been restricted to the eastern portion of the state. The Texas Bird Records Committee recognizes only eight records in Texas since 1950. These records are well documented and supported by photographs or specimens. In addition, 43 banded American Black Ducks have been taken by hunters in Texas. Wings of 35 Texas harvested American Black Ducks have also been submitted to the USFWS as part of their waterfowl harvest survey. However, voucher records in the form of archived photographs or archived specimens are not associated with the band and wing records.

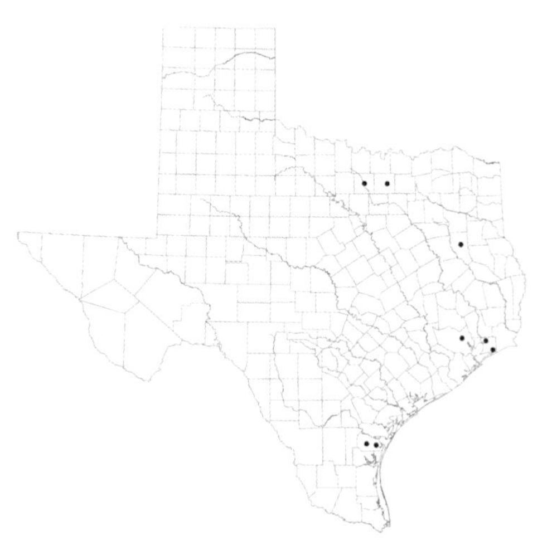

American Black Ducks breed in eastern Canada and the northeastern United States. Ecologically, they are similar to Mallards and frequently hybridize with them. They nest on the ground. Nests are found in wooded areas, in coastal marshes (above the tide line), and in and around cultivated fields. High nest densities may occur on islands. American Black Ducks occasionally nest in duck blinds, artificial nest structures, and large stick nests (for example, old heron nests). They primarily winter east of the Mississippi River, especially in wetlands along the middle and northern Atlantic coast. Wintering American Black Ducks in inland areas primarily consume vegetation, but those in coastal areas have a diet that is largely animal matter (for example, mollusks).

American Black Duck (male). *Photograph by Mark W. Lockwood, February 8, 2006, Plymouth, Mass.*

American Black Ducks are similar in size to Mallards. Both males and females have very dark bodies with lighter heads and necks. They have less buff edging on their body feathers than do Mottled Ducks; thus, they appear considerably darker. Males have yellow bills and females have light olive bills.

The American Black Duck population was estimated to number 545,000 in 2011. Their long-term population trend is declining, although their numbers are stable in parts of their breeding range. In western portions of their winter range (that is, in the Mississippi Flyway), their numbers have declined disproportionately more than in the East.

SOURCES

Stotts and Davis 1960; Bellrose 1980; Jorde and Owen 1990; Longcore et al. 2000; Lockwood and Freeman 2004; NAWMP 2004; USFWS 2011, Johnson and Garrettson 2010.

MALLARD*

Anas platyrhynchos

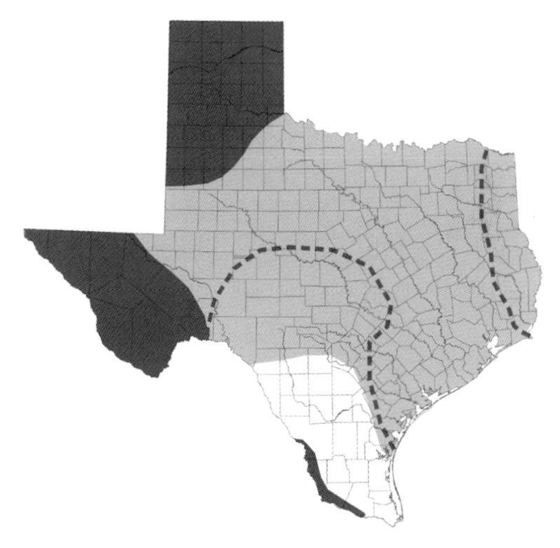

In Mallards, female body condition and production, or the number of young successfully produced during the breeding season, are correlated to the condition of wetlands used during winter and spring. This means that habitat conditions during winter and spring influence ability of females to successfully raise young. The link between wintering grounds and production reinforces the need to conserve wetlands Mallards use throughout the annual cycle.

TEXAS DISTRIBUTION

Breeding: Wild Mallards are apt to breed in low densities across most of the state. They are the most abundant breeding duck in the High Plains. In a four-year study during the 1970s, estimated brood production in the High Plains ranged from 696 to 1,528 annually in a 12-county area.

Approximately 3,070 Mallards were released near Kingsville in 1962 in an attempt to establish a local breeding population. Only a few were documented nesting in the years that followed, and brood surveys suggested production was very low. Six years after release, the few that remained were associated with farm ponds where supple-

*Mexican Ducks, a distinct subpopulation of Mallards, are addressed in their own account, immediately following.

Mallard (female and male). *Photograph by Raymond S. Matlack, October 25, 2005, Amarillo, Potter County, Texas.*

mental feeding occurred year-round. Mallard releases have also been attempted in other parts of the state, including Fannin County.

Migration: The start of fall migration is sometimes hard to gauge in the High Plains, because low densities of Mallards occur year-round. Noticeable increases are evident in this area during early October, and fall migration peaks during early November. In the northern Rolling Plains and Post Oak Savannah–Blackland Prairies, migrants begin arriving in early September. In the Coastal Prairies, fall migrants do not appear in large number until November, and their abundance peaks in December and January. Numbers in the Coastal Prairies begin declining in February. Spring migration in the High Plains peaks in late February. Migrants linger in the Rolling Plains and Post Oak Savannah–Blackland Prairies until mid-May.

Winter: From 2000 to 2008, Mallards averaged 710,691 during the Texas Mid-winter Waterfowl Survey. The vast majority of Mallards that winter in Texas are found in the High Plains, Post Oak Savannah–Blackland Prairies, and Rolling Plains (TPWD unpublished). Locally substantial to small numbers occur throughout the rest of the state.

TEXAS HARVEST

From 1999 to 2006, Mallard harvest in Texas averaged 191,120 annually, which was about 4 percent of their annual US harvest.

LONGEVITY

The life span of adult males and adult females averages 2.1 and 1.6 years, respectively. The longevity record for a wild Mallard is 27 years, seven months.

POPULATION STATUS

Mallards are the most abundant duck in North America. In 2011 the estimated population was 9.2 million. From 1955 to 2011 their population cycled up and down with wetland conditions, ranging from a low of 5 million to a high of 11.2 million. Mallard numbers are currently above the North American Waterfowl Management Plan objective of 8.2 million.

DIET

Invertebrates accounted for approximately 72 percent of the diet of breeding Mallards. In contrast, seeds and plant material dominate the diet of migrating and wintering Mallards. Food items consumed by wintering Mallards in Nacogdoches County consisted of 94 percent acorns; in Cherokee County they consisted of 62 percent acorns and 31 percent seeds. Late winter and spring migrants collected in stock ponds in Delta, Hopkins, Hunt, and Lamar Counties consumed 84 percent seeds. Migrating and wintering Mallards in the High Plains typically made both morning and evening field feeding flights, and waste corn accounted for 90 percent of their winter diet. However, nocturnal observations of ducks foraging in playa wetlands, which were not factored into the previous estimate, suggested they also forage extensively on natural foods.

RANGE AND HABITATS

Breeding: Mallards breed in Europe, Asia, and North America and have been introduced elsewhere. In North America they breed from northwestern Texas to Alaska and Nova Scotia. They also nest in California and the Intermountain West. The highest nesting concentrations are found in the Prairie Pothole Region. Breeding pairs settle in on seasonal and semipermanent prairie wetlands. They frequently nest in emergent vegetation zones of prairie wetlands, in man-made nesting structures, and occasionally in trees (snags, crotches, or hollow stumps). However, the vast majority of Mallards nest in upland habitats. Upland nests are typically located in grasslands, grassy meadows, vegetated roadsides, and croplands that offer abundant herbaceous cover, such as winter wheat and alfalfa. Nests are often concealed from above by vegetation. Upland nests are typically located near wetlands, although in drought years many are located over 100 yards from water.

Migration and Winter: Migrating and wintering Mallards occur throughout North America. The northward distribution of wintering Mallards is primarily limited by the availability of unfrozen water. They have a strong affinity for bottomland hardwoods but are also found on rivers, beaver ponds, freshwater marshes, cypress swamps, reservoirs, stock ponds, sheet water, rainwater basins, playa wetlands, and other wetlands. Along the coast they use freshwater deltas of major rivers and fresh, intermediate, and brackish marshes. Mallards tend to use water less than 16 inches deep. They regularly forage in corn, sorghum, and rice fields.

REPRODUCTION

Pair Bonds: Mallards are seasonally monogamous. Pair bonds are formed in fall; in fact, over 50 percent of females were paired when they arrived on their southernmost wintering area. Females select new mates each year, and males follow their mates north in spring. Females often return to the same breeding area in successive years. Pair bonds dissolve during incubation.

Nesting: Only females construct nests and incubate eggs. Females line nests with down and plant litter, which is obtained from the immediate vicinity of the nest site. Eggs are laid at a rate of one per day. Clutch size averages nine eggs. Effective incubation actually begins well before the last egg is laid. Females take brief incubation breaks in the morning and evening. Their incubation period is about 28 days. If their nest is lost, most females attempt to renest. Once a clutch hatches, it is extremely rare for females to make subsequent nesting attempts in the same season. Nest parasitism is uncommon unless nest densities are high.

Ducklings: Nests began hatching in the High Plains during early April, and hatching peaked during early June. All eggs within a clutch hatch within a span of about 10 hours. Only females care for young. Females lead ducklings permanently away from the nest shortly after they hatch. Females brood ducklings, give them vocal cues associated with predator avoidance, and frequently lead them overland in search of wetlands offering better foraging opportunities. Broods have been known to occupy more than 10 wetlands before fledging. Females generally stay with ducklings until they fledge. Ducklings fledge at 49–60 days.

APPEARANCE

Breeding: Mallards exhibit strong sexual dimorphism. Males have a dark green head, yellow or greenish-yellow bill, white neck ring, and chestnut-brown chest. Their back is brownish gray, their rump is black with white outer tail feathers, and their sides are pale. The black central tail feathers of males curve upward, sometimes forming complete curls. Females in breeding plumage are brownish with a broken, streaked appearance (sometimes described as mottled brown). Their body feathers are brown with white, tawny, gray, and black highlights. Females have an orange bill with black splotches. Mallard weight varies with season, sex, and age. Adult males and females average 2.8 and 2.4 pounds, respectively.

Nonbreeding: Males and females in nonbreeding plumage appear quite similar. They have a mottled or streaky brown appearance, similar to that of breeding females. However, females have an orange bill with black splotches, and males have a bill that is yellow or olive.

SOURCES

INTRODUCTION: Heitmeyer and Fredrickson 1981; Devries et al. 2008. TEXAS DISTRIBUTION: Kiel 1970; Traweek 1978; Pulich 1988; Esslinger and Wilson 2001; Seyffert 2001; White 2002; Lockwood and Freeman 2004; Baar et al. 2008; USFWS 2008; Johnson et al. 2010. TEXAS HARVEST: Kruse 2007. LONGEVITY: Bellrose and Holm 1994; Lutmerding and Love 2011. POPULATION STATUS: NAWMP 2004; USFWS 2011. DIET: Swanson et al. 1979, 1985; Allen 1980; Moore 1980; Baldassarre and Bolen 1984; Whyte and Bolen 1985; Anderson and Smith 1999; Kraai 2003; Miller et al. 2003. RANGE AND HABITATS: Cowardin et al. 1967; Stewart and Kantrud 1973; Chabreck 1979; Bellrose 1980; Heitmeyer and Fredrickson 1981; Heitmeyer and Vohs 1984; Heitmeyer 1985; Whyte and Bolen 1985; Duebbert and Kantrud 1987; Frentress 1987; Weller 1988; Smith et al. 1989; Arnold et al. 1993; Ray and Higgins 1993; Johnson and Rohwer 2000; Drilling et al. 2002; Bogiatto and Karnegis 2006; Stafford et al. 2007; Baar et al. 2008. REPRODUCTION: Hochbaum 1944; Bjarvall 1968; Dzubin and Gallop 1972; Bellrose 1980; Talent et al. 1983; Greenwood et al. 1987, 1995; Heitmeyer 1988; Johnson and Grier 1988; Rhymer 1988; Rohwer and Freeman 1989; Lokemoen et al. 1990; Batt et al. 1992; Rotella and Ratti 1992; Dzus and Clark 1997; Johnson and Rohwer 1998; Evard 1999; Stafford et al. 2001; Drilling et al. 2002; Loos and Rohwer 2004. APPEARANCE: Bellrose 1980; Krapu 1981; Drilling et al. 2002.

MEXICAN DUCK (MEXICAN MALLARD)

Anas platyrhynchos diazi

Mexican Ducks were once considered a separate species (*Anas diazi*) and were even endangered due to perceived declines in the US portion of their range. After extensive hybridization with Mallards was documented in the northern portion of their range, the endangered species status was removed. Although hybridization is likely not an issue in the southerly extent of their range, they are also no longer considered to be a distinct species. Genetic evidence, which was not available at the time the species was combined with Mallards, suggests they are more closely related to Mottled Ducks than to Mallards. Most of the information in the account below, including harvest statistics, is from studies of Mexican-like ducks that occur in the northern limit of the range.

TEXAS DISTRIBUTION

Surveys in the 1970s suggested there were about 250 breeding pairs in the Trans-Pecos. These are conservative estimates, as survey methodology did not use expansion factors to account for nonsurveyed habitats. Wintering Mexican Ducks are most common in the Trans-Pecos region, but there are no estimates of their winter numbers.

POPULATION STATUS

There are approximately 55,000 Mexican Ducks, including about 5,000 in the United States. Data from Mexico suggest their population increased between 1966 and 2000.

RANGE AND HABITATS

Mexican Ducks breed in the highlands of interior Mexico, in southeastern Arizona, in southern New Mexico, locally in the Trans-Pecos, and along the Rio Grande from El Paso southeast to Hidalgo County. Breeding pairs frequent rivers, riparian areas, flooded fields, and ponds. They nest on the ground; their nests have been found in wet meadows, dense brush thickets, and thick vegetation near ponds. Nests are typically located near water. Long-term survey data from Mexico indicate migration is not likely. Nonbreeding and wintering individuals are likely nomadic, moving in response to availability of surface water. Wintering Mexican Ducks take advantage of agricultural crops, irrigation water, and depressional wetlands.

Mexican Duck (male). *Photograph by Mark W. Lockwood, November 15, 2008, Alpine, Brewster County, Texas.*

REPRODUCTION

Mexican Ducks form seasonal pair bonds. Most females are paired in January, and pair bonds dissolve during incubation. Few nests have been found. Average clutch size is likely similar to that of Mallards. Only females care for ducklings. Ducklings are found on creeks, ponds, and irrigation canals. The duration of their fledging period is unknown.

APPEARANCE

Males and females average 2.3 and 2.0 pounds, respectively. Both sexes closely resemble female Mallards but are notably darker. Females have an orange bill that is splotched with black. Males have yellow bills. In Texas, where the population consists of many hybrids, males often have greenish feathers on their forehead. Their breeding and nonbreeding plumages are similar.

SOURCES
INTRODUCTION: Aldrich and Baer 1970; USFWS 1978; McCracken et al. 2001. TEXAS DISTRIBUTION: Swepston 1979. POPULATION STATUS: Rose and Scott 1997; Pérez-Arteaga et al. 2002. RANGE AND HABITATS: Lindsey 1946; Ohlendorf and Patton 1971; Nymeyer 1975; O'Brien 1975; Hubbard 1977; Bellrose 1980; Lockwood and Freeman 2004; Pérez-Arteaga and Gaston 2004. REPRODUCTION: Lindsey 1946; O'Brien 1975; Hubbard 1977; Swepston 1979. APPEARANCE: Huey 1961; Bellrose 1980; Drilling et al. 2002.

MOTTLED DUCK

Anas fulvigula

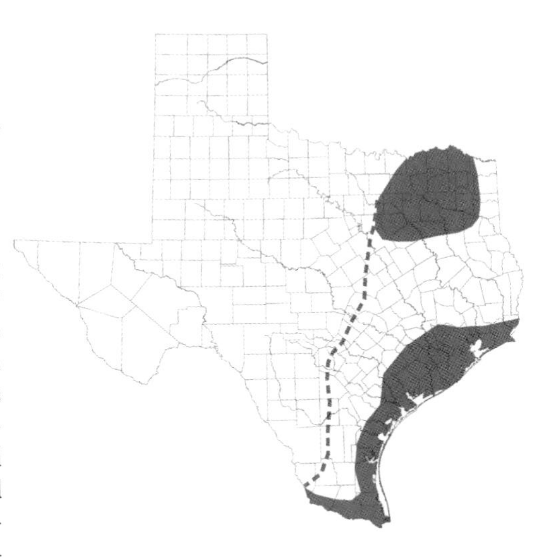

Although rates of ingestion of lead pellets (spent lead shot) by Mottled Ducks have declined since regulations were implemented that prohibited hunting waterfowl with toxic shot, their ingestion rate is still among the highest of any North American duck. Ingestion of lead pellets occurs while ducks are foraging; the shot is either picked up as grit or as a potential food source. From 1998 to 2002, approximately 14 percent of Mottled Ducks harvested on national wildlife refuges and wildlife management areas in Texas contained ingested lead pellets. Lead pellet densities that exceed 600,000 per acre have been found in some Texas coastal wetlands. These estimates are the highest reported in North American wetlands.

TEXAS DISTRIBUTION
Breeding: Mottled Ducks nest in all coastal counties and in many first-tier inland counties. They nest inland along the Rio Grande to Zapata County and inland along the central coast to Lavaca, Colorado, Austin, and Waller Counties. Their breeding density is fairly light (about one breeding pair per square mile) throughout most of their range. Their highest breeding densities occur in the coastal marshes and prairies between Galveston Bay and Sabine Lake.

Small numbers of Mottled Ducks are regularly documented in north central and northeast Texas. Interestingly, attempts were made to establish a breeding population in northeast Texas during the mid-1970s. Although it is doubtful that this propagation attempt is the source of today's northeastern Mottled Ducks, it cannot be ruled out.

Winter: From 2000 to 2008, Mottled Ducks averaged 26,084 during Texas Midwinter Waterfowl Surveys. They are most abundant in the Coastal Prairies, although

Mottled Duck (male). *Photograph by Tim Cooper, April 1997, Laguna Atascosa National Wildlife Refuge, Cameron County, Texas.*

they are regularly observed in the South Texas Brush Country and Coastal Sand Plain (TPWD unpublished).

TEXAS HARVEST

From 1999 to 2006, Mottled Duck harvest in Texas averaged 11,399 annually. This was 20 percent of their estimated annual US harvest.

LONGEVITY

The longevity record for a wild Mottled Duck is 13 years, five months. It was banded in August 1963 in Jefferson County and recovered in November 1975 at Sea Rim State Park, which is also in Jefferson County.

POPULATION STATUS

Mottled Duck population estimates and population trends vary across surveys. Estimates derived from banding data suggest the fall population in Texas and Louisiana hovers at about 630,000. In the mid-1990s the estimated spring population in Texas was 220,000. Christmas Bird Counts from 1980 to 2007 suggest the Texas population is stable. However, trend data from several sources, including USFWS harvest surveys, the Breeding Bird Survey, and midwinter waterfowl surveys, suggest Mottled Ducks are declining in Texas. Breeding pair surveys conducted on national wildlife refuges in Texas also suggest sharp declines, particularly since 1995.

Hybridization with feral Mallards (semiwild or park ducks with origins from domestic stock) is a major conservation concern. Mottled Ducks in Florida have been so overwhelmed by the introgression of Mallard genes that their continued recognition as Mottled Ducks is in jeopardy. A winter 1998–99 survey of hunter-killed Mottled Ducks in Texas suggested 6 percent have hybrid wing characteristics. The 2007 publication *Gulf Coast Joint Venture: Mottled Duck Conservation Plan* called for efforts to minimize interactions between Mottled Ducks and feral, resident, and released Mallards.

DIET

Breeding Mottled Ducks forage heavily on invertebrates, including amphipods, crayfish, dragonfly nymphs, midge larvae, and snails. Estimates of invertebrate consumption during fall and winter vary greatly, ranging from 1 percent to 40 percent. Seeds are heavily consumed during fall and winter, particularly those of rice, barnyard grass, smartweed, and widgeongrass.

RANGE AND HABITATS

Breeding: Mottled Ducks are found in peninsular Florida and along the northwestern Gulf coast from Alabama west and south to Veracruz. Small numbers breed in northeastern Louisiana and in north central and northeastern Texas. Breeding pairs use depressional wetlands, coastal ponds, irrigation canals, rice fields, and coastal marshes. Pairs tend to select ponds located in fresh marsh over those located in higher salinity zones, although they often use ponds located in intermediate and brackish marshes. Except for a few isolated areas, they nest at low densities throughout their range. They are upland nesters. Nests are typically found in pastures, coastal prairie, grass-covered ridges, and idle agricultural fields. Heavily grazed pastures are seldom used. Occasionally, nests are suspended several inches above ground in matted stands of vegetation. Nests are almost always concealed from above by vegetation.

Migration and Winter: Mottled Ducks are nonmigratory and relatively sedentary. The majority of banded Mottled Ducks are recovered within 60 miles of their banding location (B. C. Wilson, USFWS, personal communication). Perhaps related to rice harvest, exceptional concentrations are sometimes observed in the fall, suggesting localized or regional movements. Habitats used during winter are similar to breeding habitats, although brackish and salt marshes may be used to a greater degree.

REPRODUCTION

Pair Bonds: Pair bonds last for only part of the year. Pair bond formation begins in August, and approximately 83 percent of females are paired by September. Females select their mates. Due to their sedentary nature, it has been suggested that females may re-pair with the same male in subsequent breeding seasons. However, there is no evidence to support this idea. Pair bonds dissolve during incubation.

Nesting: A relatively low percentage (31–77 percent) of females attempt to nest each year. Both nest construction and incubation are solely activities of the female. Females line their nest bowl with down and plant litter, which they obtain from the vicinity of the nest site. Most nests are initiated during February, March, April, and May, but the exact timing of nesting may vary as a result of rainfall. Eggs are laid at a rate of 1 per day. Clutch size averages 9–10 eggs. Females take brief incubation breaks in the morning and evening. Their incubation period is about 26 days. Females disturbed at the nest may attempt to lead predators away by feigning injury. Up to five nesting attempts have been documented by a single female. Once a clutch hatches, females will not attempt to nest again, even if ducklings are lost. Nest parasitism by Mottled Ducks is uncommon unless nest densities are high.

Ducklings: Only females care for young. Females lead ducklings permanently away from the nest within 23 hours after they hatch. Females brood ducklings on both hot and cold days, lead ducklings to food, and make verbal cues associated with predator avoidance. Duckling survival decreases when wetland salinities exceed 9 ppt, and they are intolerant of salinities over 12 ppt. Ducklings fledge at 63–70 days.

APPEARANCE

Breeding: Although Mottled Ducks are sexually dimorphic, differences between males and females are subtle. Both have a mottled appearance, but the buff edging on female body feathers is slightly broader than that of males, giving them a lighter appearance. The head and neck of both males and females is tawny and considerably lighter than the rest of their body. Both sexes have an eye stripe. Males have yellow to olive bills, and females have orange bills with black splotches. Immature females that have recently fledged may have an olive bill, but it typically has some black spotting. Adult males and females average 2.4 and 2.1 pounds, respectively.

Nonbreeding: This plumage is similar to their breeding plumage.

SOURCES

INTRODUCTION: Fisher et al. 1986; Merendino et al. 2005. TEXAS DISTRIBUTION: Stutzenbaker 1988, Benson and Arnold 2003; Lockwood and Freeman 2004; USFWS 2008. TEXAS HARVEST: Kruse 2007. LONGEVITY: Clapp et al. 1982. POPULATION STATUS: Anderson et al. 1998; Ballard et al. 2001; McCracken et al. 2001, NAWMP 2004; Sauer et al. 2008, Wilson 2007; National Audubon Society 2010. DIET: Stutzenbaker 1988; Moorman and Gray 1994. RANGE AND HABITATS: Stutzenbaker 1988; Moorman and Gray 1994; Holbrook et al. 2000; Durham and Afton 2003; Haukos et al. 2010. REPRODUCTION: Engeling 1950; Stutzenbaker 1988; Weller 1988; Moorman et al. 1991; Batt et al. 1992; Grand 1992; Moorman and Gray 1994; Johnson et al. 1996, 2002; Holbrook et al. 2000; Durham and Afton 2003, 2006; Finger et al. 2003; Rigby 2008. APPEARANCE: Stutzenbaker 1988; Moorman and Gray 1994; Haukos et al. 2001.

BLUE-WINGED TEAL

Anas discors

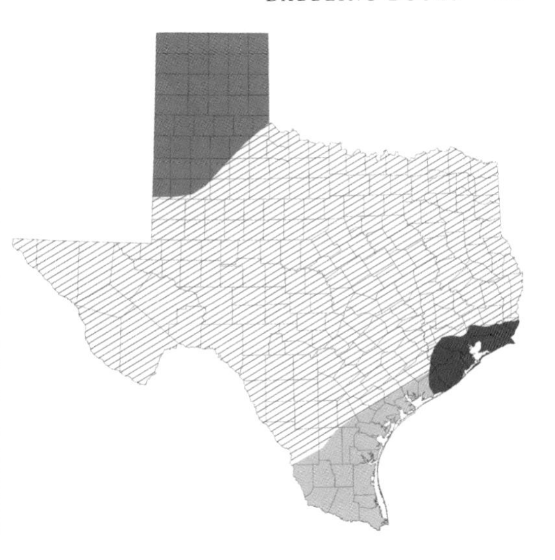

Judging from their behavior, one would guess Blue-winged Teal are not fans of cold weather. Adult males begin departing their breeding grounds in late July or early August and are the first ducks to arrive in Texas in large numbers. They are followed shortly thereafter by females and young of the year. Most Blue-winged Teal migrate through Texas and winter south of the continental United States.

TEXAS DISTRIBUTION

Breeding: Low numbers of Blue-winged Teal regularly nest in the Coastal Prairies and High Plains. Occasional nesting may occur in other areas of the state as well. In a four-year study conducted during the 1970s, brood abundance estimates in 12 High Plains counties ranged from 30 to 175 annually. They are the second most abundant breeding duck in the High Plains. In 1958, the year following Hurricane Beulah, more than 100 broods were counted on the King Ranch; when normal conditions returned the subsequent year, only two broods were counted.

Migration: Blue-winged Teal likely migrate through all Texas counties. Fall abundance in the High Plains peaks in late August and declines steadily through October; they are largely absent from this region by November. In the Rolling Plains, Post Oak Savannah–Blackland Prairies, and Pineywoods, fall migrants may be present as early as late July. Fall abundance in these regions peaks in September and then declines through November. Fall abundance peaks in the Coastal Prairies during September and drops precipitously by October. After fall migrants depart, abundance in the Coastal Prairies remains relatively flat until spring migrants arrive in March. Spring migrants start arriving in the Rolling Plains, Post Oak Savannah–Blackland Prairies, and Pineywoods in February or March and peak in April. Spring migrants arrive in the High Plains during April. Interestingly, spring abundance in the High Plains pales in comparison to fall abundance; the difference is approximately thirtyfold.

Winter: From 2000 to 2008, Blue-winged Teal averaged 96,587 during the Texas Mid-winter Waterfowl Survey. This was likely less than 2 percent of their total population (continental population) in most years. Blue-winged Teal wintering in Texas are primarily found in the Coastal Prairies, South Texas Brush Country, and Coastal Sand Plain (TPWD unpublished).

Blue-winged Teal (male and female). *Photograph by Raymond S. Matlack, December 27, 2005, near Port Aransas, Nueces County, Texas.*

TEXAS HARVEST

Harvest estimates for Blue-winged Teal and Cinnamon Teal are combined nationwide. In Texas, however, it is probably safe to assume the vast majority of the harvest is Blue-winged Teal. From 1999 to 2006, harvest in Texas averaged 173,095 annually, which was about 18 percent of their annual US harvest.

LONGEVITY

The life span of adult males and females averages 1.9 and 1.5 years, respectively. The longevity record for a wild Blue-winged Teal is 23 years, three months.

POPULATION STATUS

In 2011 the estimated abundance of Blue-winged Teal was 8.9 million, which was the highest recorded estimate since the Waterfowl Breeding Population and Habitat Survey was initiated in 1955. They are the second most abundant duck in North America. They have been above the North American Waterfowl Management Plan's population goal of 5.3 million since 2006.

DIET

Invertebrates, seeds, and vegetative matter are consumed in varying amounts during all seasons. Invertebrates accounted for 85–97 percent of the diet of breeding Blue-winged Teal. In contrast, 73 percent of the diet of fall migrants in the High Plains was barnyard grass (millet) seed. Migrants in the High Plains also consumed waste grain when it was made available by flooding, and they occasionally participated in field feeding. Wintering Blue-winged Teal in the Coastal Prairies readily used rice fields, and cultivated rice accounted for 92 percent of their diet in Costa Rica.

RANGE AND HABITATS

Breeding: Blue-winged Teal breed in low densities throughout much of the United States and Canada. They are one of the most abundant breeding ducks in the Prairie Pothole Region. Their nests are almost always found within 165 yards of wetlands. They nest on the ground, typically in short herbaceous cover. Nest density peaks in areas with abundant shallow wetlands. When wetland densities are equivalent, their abundance is four times greater in grassland settings compared to cropland settings. Breeding pairs associate strongly with ephemeral, seasonal, and semipermanent wetlands. Their affinity for ephemeral water likely explains why they occasionally breed in large numbers outside of their core range.

Migration and Winter: Migrants occur throughout most of North America, although their densities are greatest east of the Rocky Mountains. Fall migrants commonly use rainwater basins and playa wetlands. Winter and migration habitats overlap along the Texas and Louisiana coasts. Most wintering Blue-winged Teal occur in Mexico, Central America, and South America. Fresh, intermediate, and brackish coastal marshes are heavily used during winter. They are also common on rice fields, shallow freshwater wetlands, and moist-soil management wetlands. Wetlands used are often 12 inches deep or less, although Blue-winged Teal used areas 12–35 inches in depth in San Patricio County oxbows. In more southerly portions of their winter range they utilize estuaries dominated by widgeongrass and mangroves.

REPRODUCTION

Pair Bonds: The majority of females likely form pair bonds during their northward migration. Females choose their mates. Paired males follow their mates to breeding areas. Females have little fidelity to their natal area or to areas where they have previously bred successfully. Pair bonds dissolve during late incubation.

Nesting: Only females construct nests and incubate eggs. Females line nests with down and plant litter, which is obtained at the nest site. Eggs are laid at a rate of 1 per day. Clutch size averages 10 eggs. Females spend increasing amounts of time at the nest with each egg laid. They take about three incubation recesses per day. Their incubation period lasts about 23 days. If their initial nest is unsuccessful, females renest at rates around 34 percent. They will not renest after a clutch hatches, even if ducklings are lost. Nest parasitism by Blue-winged Teal is rare.

Ducklings: All eggs in a clutch hatch within about a 4-hour span. Females lead ducklings permanently away from the nest within 24 hours after they hatch. Females provide little care other than brooding, leading young to food, and perhaps giving vocal cues associated with eating and predator avoidance. Young are capable of flight at about 40 days.

APPEARANCE

Breeding: Adult males have steel-blue heads with a large white crescent between the bill and the eye. The steel-blue plumage of the head trails down to a tan body with dense black spotting. Males have a notable white patch on each side of their body, which separates their black rump from their tan sides. The body of females is mottled brown in appearance, and they have a dark eye-line. Males have a very dark, bluish black bill. Females have a slaty bill with dark spots. True to their name, both males and females have a large, chalky blue patch on the upper portion of their wings. This patch is clearly visible in flight at all times of year. In some lighting conditions the patch may appear more white than blue. Blue-winged Teal are typically heaviest during fall migration. During fall, adult males and females average 15.1 and 14.9 ounces, respectively.

Nonbreeding: Both males and females are mottled brown in appearance.

SOURCES

INTRODUCTION: Aldrich 1949; Rohwer et al. 2002. TEXAS DISTRIBUTION: Traweek 1978, Bellrose 1980; Pulich 1988; Esslinger and Wilson 2001; White 2002; Ray et al. 2003; Baar et al. 2008; USFWS 2008, 2011. TEXAS HARVEST: Kruse 2007. LONGEVITY: Bellrose and Holm 1994; Lutmerding and Love 2011. POPULATION STATUS: Rohwer et al. 2002; NAWMP 2004; USFWS 2011. DIET: Swanson et al. 1974; Swanson and Meyer 1977; Sell 1979; Baldassarre and Bolen 1984; Botero and Rusch 1994. RANGE AND HABITATS: Stewart and Kantrud 1973; Duebbert and Lokemoen 1976; Taylor 1978; White and James 1978; Bellrose 1980; Livezey 1981; Thomas 1982; Heitmeyer and Vohs 1984; Smith et al. 1989; Thompson and Baldassarre 1991; Fischer 1998; Rohwer et al. 2002; Baar et al. 2008. REPRODUCTION: Bennett 1938; Rohwer 1985; Clark et al. 1988; Batt et al. 1992; Thompson and Baldassarre 1992; Rohwer et al. 2002; Loos and Rohwer 2004; Wells-Berlin et al. 2005. APPEARANCE: Havera 1999a; Rohwer et al. 2002.

CINNAMON TEAL

Anas cyanoptera

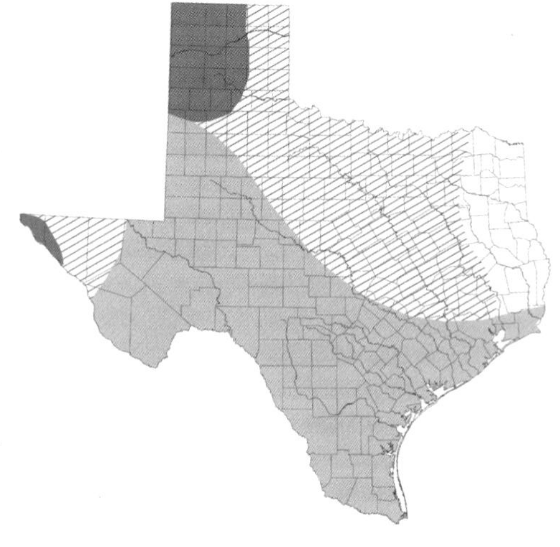

No two species of North American ducks look more alike in their nonbreeding plumage than Cinnamon Teal and Blue-winged Teal. In fact, female Cinnamon Teal and female Blue-winged Teal are nearly identical in appearance at all times of year, although Cinnamon Teal tend to have slightly more spatulate and slightly longer bills. Bills of female Cinnamon Teal are, on average, 0.2 inches longer than those of female Blue-winged

Cinnamon Teal (male). *Photograph by Greg Lasley, March 2, 2009, Austin, Travis County, Texas.*

Teal but can be up to 0.4 inches longer. Unfortunately, bill length is not a diagnostic trait, as there is a slight overlap in the range of bill measurements for the two species.

TEXAS DISTRIBUTION

Breeding: Small numbers of Cinnamon Teal regularly nest in the High Plains and in El Paso County. There are also breeding records for Bexar and Colorado Counties.
Migration and Winter: Cinnamon Teal migrate through the western three-fourths of the state. In the Rolling Plains and Post Oak Savannah–Blackland Prairies, most reports of migrants are from February. They are locally common in the Trans-Pecos and local to uncommon during winter throughout the Coastal Prairies. They winter in the High Plains irregularly. Rarely, they winter inland in other parts of Texas. There are no estimates of their wintering numbers.

POPULATION STATUS

The Breeding Bird Survey suggests a slight negative trend from 1967 to 2006. There is no population estimate for Cinnamon Teal.

DIET

Cinnamon Teal forage on invertebrates, seeds, and vegetative parts of aquatic plants. In Tulare Lake Basin, California, prenesting males consumed invertebrates. In Arizona, invertebrates accounted for 75 percent of the diet of laying females. In Sinaloa, 67 percent of the diet of wintering Cinnamon Teal was plant material (mainly seeds).

RANGE AND HABITATS

Breeding: Cinnamon Teal breed in low densities from southwestern Canada south into western Mexico. They reach their highest breeding densities around Great Salt Lake, Utah. They also breed in South America, but breeding ranges of South American and North American populations do not overlap. Breeding pairs are found on freshwater wetlands, including stock ponds, playa wetlands, marshes, riparian wetlands, montane wetlands, and other seasonal and semipermanent wetlands. They nest on the ground and generally near water. Nests are located in short, dense herbaceous cover such as saltgrass, spikerush, and western wheatgrass. Nests are frequently located in dry wetlands, and Cinnamon Teal readily nest on islands.

Migration and Winter: Habitats used by migrants include freshwater marshes, playa wetlands, riparian wetlands, moist-soil management wetlands, and other moderately to heavily vegetated shallow wetlands. Most Cinnamon Teal winter in Mexico, where coastal wetlands and rice fields are commonly used.

REPRODUCTION

Pair Bonds: Pair bonds are likely formed during spring migration. Females choose their mates, which subsequently follow them to breeding areas. Pair bonds dissolve during incubation.

Nesting: Nest construction and incubation are activities of the female. Females line nests with down and plant litter, which is obtained from the vicinity of the nest site. Eggs are laid at a rate of 1 per day. Clutch size averages 10 eggs. Females take multiple incubation breaks per day. Their incubation period is 21–25 days. If their initial nest attempt is unsuccessful, about 13 percent of females renest. During late incubation, females will feign injury to lead predators away from the nest. They occasionally parasitize nests of other ducks.

Ducklings: Only females care for young. Females lead ducklings permanently away from the nest shortly after they hatch, actively brood ducklings during their first week, and are aggressive toward other birds that approach their young. Young fledge at about 49 days.

APPEARANCE

Breeding: Adult males have a cinnamon red body color. The top (crown) and back of their head are dark brown to black. Their belly ranges from reddish brown to black. The long feathers on their back are black, tawny, and cinnamon. Adult females are mottled brown in appearance. Both male and female Cinnamon Teal have a large, chalky blue patch on the upper portion of their wings. Males have a glossy black bill. Females have a slaty bill with black spotting near the edges. During the breeding season, adult males and females average 13.5 and 13.1 ounces, respectively.

Nonbreeding: Both males and females are mottled brown in appearance.

SOURCES

INTRODUCTION: Stark 1978; Lokemoen and Sharp 1981; Gammonley 1996; Rohwer et al. 2002. TEXAS DISTRIBUTION: Traweek 1978; Pulich 1988; Lockwood and Freeman 2004; Baar et al. 2008. POPULATION STATUS: Gammonley 1996; Sauer et al. 2008. DIET: Migoya and Baldassarre 1993; Hohman and Ankney 1994; Gammonley 1995, 1996. RANGE AND HABITATS: Gray and Schultze 1977; Bellrose 1980; Weller 1988; Smith et al. 1989; Gammonley 1996; Elphick and Oring 1998; Gammonley and Fredrickson 1998. REPRODUCTION: Miller and Collins 1954; Hunt and Anderson 1966; Joyner 1973; Bellrose 1980; Hohman 1991; Gammonley 1996. APPEARANCE: Stark 1978; Bellrose 1980; Gammonley 1996.

NORTHERN SHOVELER

Anas clypeata

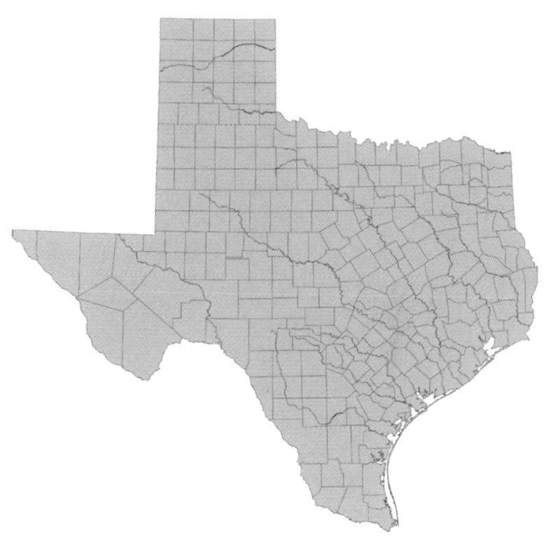

The large, broad bill of Northern Shovelers not only sets them apart from other ducks in appearance, but also in diet. They specialize on planktonic and benthic organisms, which they filter out of the water and substrate with dense comblike projections (lamellae) that grow along the sides of their bill. They regularly forage on small food items that are seldom recorded in the diets of other ducks, such as unicellular forms of plankton.

TEXAS DISTRIBUTION

Breeding: Reports from early naturalists suggest Northern Shovelers may have bred throughout much of the state, including the Coastal Prairies. Although adults are uncommon to locally common throughout the state during summer, they are now rare breeders in Texas. With the exception of Bastrop and Bexar Counties, most recent records are from the High Plains and Trans-Pecos.

Migration: Northern Shovelers migrate through all Texas counties. In the High Plains migrants arrive in August, and their numbers increase through early October. Their numbers decline during winter, although some remain during winter months. In the northern portions of the Rolling Plains, Post Oak Savannah–Blackland Prairies, and Pineywoods, the first migrants arrive in August or early September. Migrants arrive in the Coastal Prairies and South Texas Brush Country in September and October; their numbers do not peak in these areas until February or March. Numbers in the Coastal Prairies begin declining during April. In the High Plains, spring migrants begin increasing in early February and remain common through May. Most have left the Rolling Plains, Post Oak Savannah–Blackland Prairies, and Pineywoods by late May.

Northern Shoveler (male). *Photograph by Mark W. Lockwood, January 17, 2009, Midland, Midland County, Texas.*

Northern Shoveler (female). *Photograph by Mark W. Lockwood, April 2, 2009, Balmorhea Lake, Reeves County, Texas.*

Winter: From 2000 to 2008, Northern Shovelers averaged 128,680 during the Texas Mid-winter Waterfowl Survey. Most Northern Shovelers that winter in Texas occur in the Coastal Prairies, although notable numbers occur in the South Texas Brush Country, Post Oak Savannah–Blackland Prairies, Rolling Plains, and Coastal Sand Plain (TPWD unpublished). Small numbers winter throughout the rest of the state.

TEXAS HARVEST

From 1999 to 2006, Northern Shoveler harvest in Texas averaged 61,516 annually. This was approximately 12 percent of their annual US harvest.

POPULATION STATUS

In 2011 the estimated abundance of Northern Shovelers was 4.6 million, which was the highest estimate since the Waterfowl Breeding Population and Habitat Survey was initiated in 1955. Northern Shovelers have been above the North American Waterfowl Management Plan's population goal of 2.1 million since 1994.

DIET

Northern Shovelers consume invertebrates (for example, snails, zooplankton, and insects) throughout the year. In Manitoba, invertebrates accounted for 93 percent of the diet of laying females. Northern Shovelers foraging in freshwater wetlands in San Patricio County, Texas, consumed 50 percent animal matter and 50 percent plant material. Water fleas and snails were the primary invertebrates consumed, and coontail was the most common plant consumed. Northern Shovelers foraging in saltwater wetlands in Refugio County consumed 80 percent animal and 20 percent plant material. Unicellular foraminifera, tiny seed shrimp, and snails were the primary invertebrates consumed; widgeongrass was the most common plant consumed.

RANGE AND HABITATS

Breeding: Northern Shovelers breed in North America, Europe, and Asia. They nest in Alaska and in small numbers throughout much of the western United States and Canada. They breed locally in eastern North America. The core of their North American breeding range is the Prairie Pothole Region. Breeding pairs settle in on shallow marshes and wetlands that are characterized by submersed aquatic vegetation and high invertebrate densities. All nests are found in uplands; nests are typically located in short herbaceous cover, including meadows and hayfields. Small areas of herbaceous cover associated with rock piles and fencerows are often used for nesting. Most nests are located within 200 feet of wetlands.

Migration and Winter: Migrants occur throughout North America but are most common west of the Mississippi River. In the United States, Northern Shovelers winter in the West, throughout the Southeast, and along the East Coast. They also winter in the Caribbean and in Mexico and occur sporadically south to Colombia. During migration and winter they commonly use saline lakes and shallow freshwater wetlands. During winter they also heavily use wetlands with abundant submersed aquatic vegetation and high invertebrate densities, such as coastal marshes and mudflats. They readily use wetlands associated with municipal systems, such as treatment wetlands. They commonly use rice fields, but they do not participate in dryland field feeding.

REPRODUCTION

Pair Bonds: In North Carolina, pair bond formation began in November, and most females were paired in February. Females choose their mates. Males follow their mates north in spring. Adult females returned to familiar breeding areas at rates ranging from 15 percent to 42 percent. Pair bonds dissolve during incubation.
Nesting: Only females construct nests and incubate eggs. They line nests with down and plant litter, which is obtained at the nest site. Eggs are laid at a rate of 1 per day. Clutch size is typically 9–11 eggs. Females take multiple incubation recesses each day. Females occasionally renest if their first attempt is unsuccessful. Once a clutch hatches, subsequent nesting attempts are not made even if ducklings are lost. The incubation period is about 24 days. During late incubation females may perform distraction displays to lead predators away from the nest. They rarely parasitize nests of other ducks.
Ducklings: Hatching is largely synchronized. Females lead their ducklings permanently away from the nest shortly after they hatch. They also brood ducklings, lead them to food, and lead them overland to new wetlands. Ducklings fledge at about 50 days.

APPEARANCE

Breeding: Adult male Northern Shovelers have an iridescent green head and neck. Their breast is white and their sides and belly are reddish. Their back is dark and their rump is black. A notable white patch separates their reddish sides from their black rump. During winter males commonly have patchy heads and patchy breasts, as they are slow to acquire their breeding plumage. Adult females have a mottled brown appearance consisting of light tan feathers edged with brown. Adult males reach their peak weight during fall, averaging 1.4 pounds. Adult females reach their peak weight while laying, also averaging 1.4 pounds.

The most noticeable feature of Northern Shovelers is their large, broad bill, which is longer than their head, ranging in length from 2.5 to 2.75 inches. The bill broadens and flattens near the tip, where it averages 1.25 inches in width. Males have black bills. Females have olive green bills with yellow edging, and they may also have small black dots on their bills.
Nonbreeding: Both males and females are mottled brown in appearance, similar to breeding females.

SOURCES

INTRODUCTION: DuBowy 1985; Ankney and Afton 1988; Migoya and Baldassarre 1993; Tietje and Teer 1996. TEXAS DISTRIBUTION: Bellrose 1980; Pulich 1988; Seyffert 2001; White 2002; Benson and Arnold 2003; Ray et al. 2003; Lockwood and Freeman 2004; Eubanks et al. 2006; Baar et al. 2008; USFWS 2008; Johnson et al. 2010. TEXAS HARVEST: Kruse 2007. POPULATION STATUS: NAWMP 2004; USFWS 2011. DIET: Ankney and Afton 1988; Tietje and Teer 1996. RANGE AND HABITATS: Lokemoen 1973; Chabreck 1979; Bellrose 1980; Greenwood et al. 1995; DuBowy 1996; Tietje and Teer 1996; Elphick and Oring 1998. REPRODUCTION: Johnsgard 1975; Afton 1980; Hepp and Hair 1983;

Lokemoen 1991; Batt et al. 1992; Kantrud 1993; MacCluskie and Sedinger 1999; Wells-Berlin et al. 2005. APPEARANCE: Bellrose 1980; DuBowy 1996.

WHITE-CHEEKED PINTAIL

Anas bahamensis

The only documented record of a White-cheeked Pintail in Texas is from Laguna Atascosa NWR, Cameron County. It was sighted repeatedly from November 20, 1978, through April 15, 1979. There is controversy over the provenance of this record and of all other records of this species in the United States.

The White-cheeked Pintail is native to the Caribbean, the Galapagos Islands, Central America, and South America. Although nonmigratory, it may wander during the nonbreeding season. White-cheeked Pintails are not similar in appearance to Northern Pintails. Adult males are

White-cheeked Pintail (male). *Photograph by Mark W. Lockwood, November 6, 1999, Paradise Island*

dark mottled brown. Their cheeks and upper neck are white, and they have a prominent red mark on their bill. Adult females are mottled brown with whitish cheeks and have a reddish-orange patch on their bill.

SOURCES

Ogilvie and Young 1998; Lockwood and Freeman 2004.

NORTHERN PINTAIL

Anas acuta

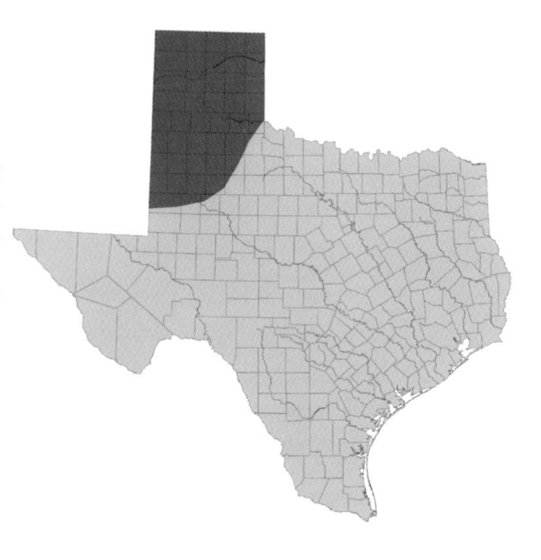

Many species have a high degree of fidelity to their nesting grounds, but pairs of Northern Pintails settle in on very shallow, ephemeral wetlands. The availability of such wetlands is highly variable from year to year; thus, Northern Pintails exhibit little breeding site fidelity. However, they are relatively faithful to wintering regions that have dependable wetlands. Approximately 65 percent of Northern Pintails banded in the Texas-Louisiana coastal region are recovered in this area in later years.

TEXAS DISTRIBUTION

Breeding: There are Northern Pintail breeding records for at least nine High Plains counties. Adults are fairly common through June, and breeding may be more common than records suggest. Most brood sightings in the High Plains have occurred during waterfowl production surveys rather than incidentally. Freshly hatched ducklings in the High Plains have been recorded in May and June. There are also breeding records for the Trans-Pecos (one), Coastal Prairies (one), South Texas Brush Country (one), and Rolling Plains (two).

Migration: Northern Pintails migrate through all Texas counties. In the High Plains migrants arrive in early August, and their numbers peak during November. Many remain in the High Plains throughout winter. In the northern portions of the Pineywoods and Post Oak Savannah–Blackland Prairies, the first migrants arrive in August, and many remain through winter. Migrants arrive in the Coastal Prairies and South Texas Brush Country in September, and their numbers peak in December or January; numbers in these two regions drop sharply in February, and the bulk are gone by March. In the High Plains spring migrants begin increasing dramatically in early February and peak in late February. Late February abundance in the High Plains far exceeds both fall and winter abundance. Most have left Texas by late April.

Northern Pintail (male). *Photograph by Trey Barron, December 27, 2007, Canyon, Randall County, Texas.*

Northern Pintail (female). *Photograph by Tim Cooper, December 8, 2004, South Padre Island, Cameron County, Texas.*

Winter: From 2000 to 2008, Northern Pintails averaged 603,647 during the Texas Mid-winter Waterfowl Survey. Most are found in the Coastal Prairies and High Plains, although they are also common in the Rolling Plains and South Texas Brush Country (TPWD unpublished). Small numbers may winter throughout the rest of the state.

TEXAS HARVEST

From 1999 to 2006, Northern Pintail harvest in Texas averaged 46,256 annually. This was approximately 11 percent of their annual US harvest.

LONGEVITY

The average life span of adult males and females is 3.0 years and 2.0 years, respectively. The longevity record for a wild Northern Pintail is 22 years, three months.

POPULATION STATUS

Although common, Northern Pintails have suffered a population decline. In 2011 their estimated abundance was 4.4 million. From 1955 to 2011 their population fluctuated from 10.4 million to 1.8 million. They have been below the North American Waterfowl Management Plan's population goal of 5.6 million since 1975.

DIET

Breeding females in North Dakota foraged heavily (77 percent) on invertebrates, particularly when laying; the primary invertebrates consumed are midge larvae and snails. Similarly, vegetation accounted for 70 percent of the diet of breeding males. In the High Plains they consumed waste grains (primarily corn) during migration and winter; they also foraged extensively in playa wetlands, consuming seeds and invertebrates. Shoalgrass (foliage and rhizomes) accounted for over 50 percent of the vegetation consumed in the Laguna Madre of Texas; widgeongrass (seeds), amphipods (*Hyalella* spp.), dwarf surf clams, and marine gastropods were also consumed. Rice is heavily consumed in the Coastal Prairies.

RANGE AND HABITATS

Breeding: Northern Pintails breed in North America, northern Europe, and northern Asia. In North America they breed throughout much of the western two-thirds of the United States and Canada, in wetlands associated with the Great Lakes, and in coastal regions of eastern Canada. Nesting in the southwestern United States is highly localized. In the Prairie Pothole Region, pairs are associated with seasonal and semipermanent wetlands. They are upland nesters, and nests are typically located in short perennial grasses. However, in parts of the Prairie Pothole Region, 34–57 percent of nests may be in crop stubble.

Migration and Winter: Northern Pintails migrate throughout North America. They winter in southeastern Alaska, along the west coast of Canada, and throughout most of the continental United States. They also winter in the Caribbean and Mexico and occur sporadically south to Colombia. Wintering and migrating Northern Pintails use playa wetlands, rainwater basins, managed wetlands, rice fields, flooded and dry cropland, freshwater ponds, mudflats, coastal wetlands, estuaries, and bays. Freshwater ponds, located just inland, are regularly used by Northern Pintails foraging in hypersaline areas along the Texas coast. In the Coastal Prairies, their use of saltwater marshes and bays may be increasing because of declines in both rice acreage and freshwater wetlands.

REPRODUCTION

Pair Bonds: Pair bonds are seasonal. In the Coastal Prairies, 87 percent of females were paired in December and January. In contrast, less than 3 percent of females in Yucatán were paired in February, just prior to spring migration. Females choose their mates, which follow them to breeding areas. Pair bonds dissolve during early incubation.

Nesting: Only females construct nests and incubate eggs. They line nests with down and plant litter, which they obtain from the immediate vicinity of the nest. Eggs are laid at a rate of one per day. Clutch size is typically seven or eight eggs. Females take multiple incubation breaks per day. Renesting is common, but it is doubtful more than three attempts are made per season. Once a clutch hatches, further nesting attempts are typically not made, even if ducklings are lost. However, two of six females were induced to renest when their ducklings were taken shortly after hatching but before leaving the nest. Incubation period of Northern Pintails is 22–24 days. They rarely parasitize other nests.

Ducklings: Females lead ducklings permanently away from the nest within 24 hours after they hatch. Females brood ducklings, lead them to food, lead them away from predators, and may feign injury to lure predators away. Female Northern Pintails may even behave aggressively toward some predators; there is an observation of one injuring a Franklin's Gull. Ducklings fledge at 42–57 days.

APPEARANCE

Breeding: Northern Pintails have a long, slender appearance. Adult males have a chocolate brown head and a white neck. The white on the neck extends upward in a tapered, arching line on each side of the head. Their back and sides are gray; some of their back feathers are long, with black centers and pale edging. Their breast and belly are white. Their rump is black but is separated from their gray sides by a creamy patch. Their two central tail feathers are black, narrow, and elongated (approximately 8.7 inches). Females have a mottled brown body, but their head and neck are notably lighter (light brown to tan) than their back and breast. Males have black bills with bluish stripes on the sides. Females have slaty bills with faint bluish stripes on the sides. Their weight fluctuates seasonally. In California, adult males and females average 2.2 and 1.8 pounds, respectively, during fall and winter.

Nonbreeding: Both males and females are mottled brown. Their slender appearance, both in flight and on water, sets them apart from other species that have mottled brown nonbreeding plumages.

SOURCES

INTRODUCTION: Hestbeck 1993. TEXAS DISTRIBUTION: Hawkins 1945; Traweek 1978; Seyffert 2001; White 2002; Ray et al. 2003; Lockwood and Freeman 2004; Baar et al. 2008; USFWS 2008; Johnson et al. 2010. TEXAS HARVEST: Kruse 2007. LONGEVITY: Bellrose and Holm 1994; Lutmerding and Love 2011. POPULATION STATUS: Austin and Miller 1995; NAWMP 2004; USFWS 2011. DIET: McMahan 1970; Krapu 1974; Baldassarre and Bolen 1984; Sheeley and Smith 1989; Cox and Afton 2000; Ballard et al. 2004. RANGE AND HABITATS: Stewart and Kantrud 1973; Bellrose 1980; Euliss and Harris 1987; Greenwood et al. 1995; Adair et al. 1996; Cox and Afton 1997, 2000; Sedinger 1997; Richkus 2002; Ballard et al. 2004; Fleskes et al. 2005. REPRODUCTION: Sowls 1955; Afton 1978; Hochbaum and Ball 1978; Duncan 1987a, 1987b; Bellrose 1980; Lokemoen 1991; Thompson and Baldassarre 1992; Hestbeck 1993; Austin and Miller 1995; Richkus 2002; Richkus et al. 2005. APPEARANCE: Bellrose 1980; Austin and Miller 1995; Havera 1999a.

GARGANEY

Anas querquedula

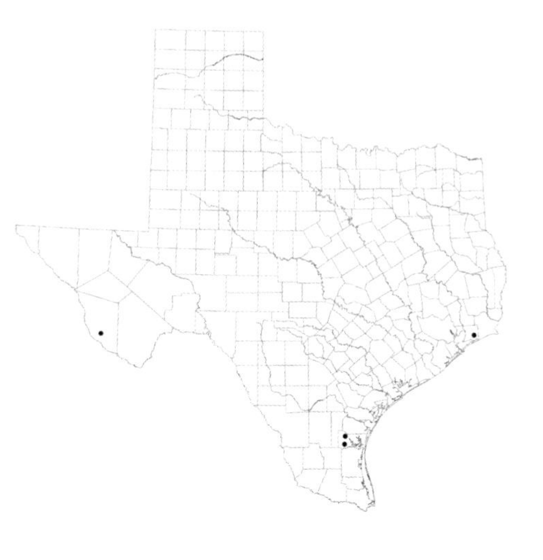

Garganey are very rare visitors to Texas, with only four records. All sightings have occurred between April 4 and May 17. The records are from Presidio, Galveston, and Kleberg (two) Counties. All sightings have involved males. Garganey breed in Europe and northern Asia, where they are found in fairly light densities. Their breeding and wintering ranges do not overlap, as most winter in southern Asia and Africa.

In nonbreeding plumage male and female Garganey are similar in appearance. They have a dark crown, dark eye-line, and dark lower facial stripe. Their lower facial stripe starts near the bottom of the upper mandible and extends partway across their cheek. Except for their face, which has a distinctly striped appearance, they are very similar to nonbreeding Blue-winged Teal in their overall appearance, color, and size. In breeding plumage males have a dark maroon brown head with a bold white stripe that starts just forward of the eye and arcs backward and down to the base of their head. During all seasons both sexes have a pale blue upper wing.

Garganey sightings in the United States are believed to involve birds that originate from northeast Asia; during fall they migrate into North America instead of southern Asia. These birds are hidden in flocks of teal while in their nonbreeding plumage. However, the distinctive breeding plumage of male Garganey is recognizable during spring migration.

SOURCES

Gooders and Boyer 1986; Lockwood and Freeman 2004.

Garganey (male). *Photograph by Mark W. Lockwood, April 30, 1994, Presidio, Presidio County, Texas.*

GREEN-WINGED TEAL

Anas crecca

Green-winged Teal are the smallest North American dabbling duck. Their diminutive size, rapid wing beats, and erratic flight often lead hunters to surmise they are the fastest duck. However, their normal flight speed is only about 30 mph. Canada Geese, Canvasbacks, Snow Geese, Long-tailed Ducks, Mallards, Northern Pintails, and Wood Ducks all have much faster flight speeds. Red-breasted Mergansers have been documented at over 80 mph, which is the fastest recorded flight speed for a duck.

TEXAS DISTRIBUTION

Breeding: There is one Green-winged Teal nesting record from the High Plains (Crosby County) and one from the Rolling Plains (Hutchinson County). Male-female pairs are somewhat common in the High Plains during May and June, and pairs may also linger into early summer in northeast Texas.

Migration: Green-winged Teal migrate through all Texas counties. In the High Plains their numbers start building in early September. Fall abundance peaks during late October and early November and drops off precipitously by late November. In the Pineywoods and Post Oak Savannah–Blackland Prairies, large numbers arrive in September and remain through winter. Numbers in the middle and upper portions of the Coastal Prairies begin increasing in September and peak in December or January. Numbers in the Coastal Sand Plain and in the lower reaches of the Coastal Prairies peak in January. They begin declining in the Coastal Prairies and Coastal Sand Plain in February, but a few are still present in late April. In the High Plains, spring numbers begin increasing in late February, peak in late March, and then decline.

Winter: From 2000 to 2008, Green-winged Teal averaged 466,293 during the Texas Mid-winter Waterfowl Survey. The vast majority are found in the Coastal Prairies, although fair numbers also winter in the High Plains (TPWD unpublished). They are locally abundant to common in the remainder of Texas.

The lone Texas record of a Eurasian Teal (*A. c. crecca*) was a male photographed at the Village Creek Drying Beds in Tarrant County in 1994.

Green-winged Teal (male). *Photograph by Raymond S. Matlack, March 5, 2006, near Port Aransas, Nueces County, Texas.*

Green-winged Teal (female). *Photograph by Greg Lasley, February 6, 2010, Austin, Travis County, Texas.*

TEXAS HARVEST

From 1999 to 2006, Green-winged Teal harvest averaged 174,127 annually. This accounted for about 11 percent of their annual US harvest.

LONGEVITY

The longevity record for a wild Green-winged Teal is 20 years, three months.

POPULATION STATUS

In 2011 their estimated abundance was 2.9 million. From 1955 to 2011, populations fluctuated from an estimated low of 722,900 to a high of 3.5 million. Green-winged Teal have been above the North American Waterfowl Management Plan's population goal of 1.9 million since 1994.

DIET

Diets of breeding Green-winged Teal are poorly studied, but invertebrates are likely important. Several studies, however, have addressed their diets during migration and winter. In the High Plains, agricultural grains (particularly corn) were consumed while field feeding, and seeds of barnyard grass and other moist-soil plants were commonly consumed by individuals foraging in playa wetlands. Invertebrates comprised 8–31 percent of their diet in the High Plains. In California's Central Valley, invertebrate consumption (particularly midge larvae) increased in late winter.

RANGE AND HABITATS

Breeding: Green-winged Teal breed in North America, Europe, and Asia. Those breeding in Europe and Asia are known as Eurasian Teal and are sometimes considered a separate species. Key breeding areas for the North American subspecies (*A. c. carolinensis*) include the aspen parklands, boreal forests, and arctic regions of Alaska and Canada. Low numbers breed in the Prairie Pothole Region and Intermountain West. Green-winged Teal are upland nesters. Their nests are often located adjacent to wetlands or streams, in grass- or sedge-dominated meadows, or in brushy cover. Concealment above and around Green-winged Teal nests is very dense compared to that for other dabbling ducks.

Migration and Winter: Migrants occur throughout North America. Green-winged Teal winter along the west coast of Canada and throughout much of the continental United States. They also winter in Mexico and in the Caribbean. Throughout their winter range they use most freshwater wetlands. These include playa wetlands, vernal pools, open areas of forested wetlands, stock ponds, reservoirs, managed wetlands, rice fields, and mudflats. They use coastal wetlands heavily, but they typically avoid salt marshes and open saltwater areas. They prefer shallow water, as they frequently walk while foraging.

REPRODUCTION

Pair Bonds: Green-winged Teal are seasonally monogamous. In Louisiana the first pair bonds were formed in January, and about 80 percent of females were paired when they departed in spring. Females choose mates from large courtship groups, which may number up to 25 males. Paired males follow their mates north in spring. Pair bonds dissolve during incubation.

Nesting: Only females construct nests and incubate eggs. They line their nest with down and plant litter, which they obtain from the immediate vicinity of the nest site. Eggs are laid at a rate of one per day, and the clutch size is eight or nine eggs. Incubating females spend about 79 percent of their time on the nest. If the initial nesting effort is unsuccessful, second attempts may occur. Their incubation period is 21–23 days.

Ducklings: Females brood ducklings, lead them to food, lead them away from predators, and perform distraction displays to lure predators away. Ducklings fledge at about 34 days.

APPEARANCE

Breeding: Adult males have rust-colored (cinnamon) heads with a slight crest and a green crescentlike line that starts just forward of the eye and arches downward to the back of the head. The rust plumage of the head ends abruptly midway down the neck. Their breast is slightly pinkish and has small black spots. Males have a vertical white stripe on each side of their breast (visible from the side), and their sides and back are gray. Each side of the rump has a cream-colored triangular patch. Females are mottled brown in appearance at all times of year. Both males and females have small, dark gray bills. Adult males and females wintering in the High Plains average 12.3 and 11.5 ounces, respectively.

Male Eurasian Teal differ slightly in appearance from male Green-winged Teal. Eurasian Teal have a horizontal white stripe separating their back and sides instead of a vertical white stripe on their sides. Female Eurasian Teal are indistinguishable from female Green-winged Teal.

Nonbreeding: Both male and female Green-winged Teal are mottled brown.

SOURCES

INTRODUCTION: Speirs 1945; Thompson 1961; Lokemoen 1967; Bellrose 1980; Bellrose and Crompton 1981; Wege and Raveling 1984; Johnson 1995. TEXAS DISTRIBUTION: Seyffert 2001; White 2002; Ray et al. 2003; Lockwood and Freeman 2004; Baar et al. 2008; USFWS 2008. TEXAS HARVEST: Kruse 2007. LONGEVITY: Lutmerding and Love 2011. POPULATION STATUS: NAWMP 2004; USFWS 2011. DIET: Rollo and Bolen 1969; Sell 1979; Baldassarre and Bolen 1984; Euliss and Harris 1987; Anderson et al. 2000a. RANGE AND HABITATS: Munro 1949; Bellrose 1980; Johnson 1995; Anderson et al. 2000a; Johnson and Rohwer 2000; Bogiatto and Karnegis 2006. REPRODUCTION: Munro 1949; Keith 1961; Koskimies and Lahti 1964; McKinney 1965; Moisan et al. 1967; Afton 1978; Rohwer and Anderson 1988; Johnson 1995; Johnson and Rohwer 1998. APPEARANCE: Bellrose 1980; Baldassarre and Bolen 1986; Johnson 1995; Havera 1999a.

CANVASBACK

Aythya valisineria

Female Canvasbacks take the lead role in mate selection, as is the norm for most North American ducks that form seasonal pair bonds. However, the role of female Canvasbacks in selecting mates is so strong that even in captivity they must have multiple males to choose from. In captivity, forced male-female groupings are almost always unsuccessful, as female Canvasbacks will reject males and even show aggression toward them. In most other species, forced male-female groupings are successful.

TEXAS DISTRIBUTION

Breeding: Canvasbacks are rare Texas nesters. In a 1988–92 study of potential breeding pairs in the High Plains, only one pair was observed. In the 1970s, 12 adults and four broods were observed during a four-year study in 12 High Plains counties. The

Canvasback (male). *Photograph by Trey Barron, February 4, 2009, Amarillo, Potter County, Texas.*

most recent evidence of breeding was a female with young in Jeff Davis County in 1986.

Migration: Canvasbacks occur throughout Texas during migration. Migrants arrive in the High Plains in November, and their presence peaks during early January. In the northern Rolling Plains, Post Oak Savannah–Blackland Prairies, and Pineywoods, the first fall migrants arrive in mid-October, and they peak in November. Fall migrants arrive in the Coastal Prairies in October, and by November they are common. In the northern Rolling Plains, Post Oak Savannah–Blackland Prairies, and Pineywoods, spring migrants peak in late February or early March. Numbers in the High Plains begin declining in late January, and most are gone by mid-March.

Winter: From 2000 to 2008, Canvasbacks averaged 18,488 during the Texas Midwinter Waterfowl Survey. They are most common in the Pineywoods, Post Oak Savannah–Blackland Prairies, Coastal Prairies, and Rolling Plains (TPWD unpublished). Canvasbacks may be locally abundant throughout the rest of the state.

TEXAS HARVEST

From 1999 to 2006, harvest in Texas averaged 8,092 annually. This is about 13 percent of their estimated annual US harvest.

POPULATION STATUS

In 2011 their estimated abundance was 691,600. From 1955 to 2011 their population fluctuated from a low of 360,200 to a high of 864,900. Their long-term population trend is stable. The North American Waterfowl Management Plan's population goal is 540,000.

DIET

Canvasbacks primarily forage along the substrate by tipping up or diving, although they forage at the surface also. When foraging on shallowly flooded tidal mudflats, they will even use their feet to excavate tubers of arrowhead and delta duck potato. In Manitoba, 78 percent of the diet of laying females consisted of invertebrates. Canvasbacks wintering in Louisiana foraged primarily on vegetation, particularly chufa, arrowhead, bulrush rhizomes, and seeds. Banana waterlily was discovered to be an important food during early studies of Canvasback natural history conducted at Lake Surprise, Chambers County. In estuarine areas, such as the Chesapeake Bay, wintering Canvasbacks foraged heavily on small clams. They take their scientific

Canvasback (female). *Photograph by Frank Rohwer, July 8, 2008, near Minnedosa, Manitoba, Canada.*

name from wildcelery (*Vallisneria americana*), a plant that was historically their most important food during migration and winter.

RANGE AND HABITATS

Breeding: The breeding range of Canvasbacks extends from the Prairie Pothole Region north-northwest to Alaska. They also breed locally in the Intermountain West. Breeding pairs settle in on semipermanent and permanent freshwater wetlands that are bordered by dense emergent vegetation. They are overwater nesters.

Migration and Winter: Migrants use wetlands throughout North America. Canvasbacks winter from southwestern British Columbia south to Baja California Sur, from coastal New England south to Florida, and in the central and southern United States. They also winter in Mexico. Canvasbacks use reservoirs, stock ponds, rainwater basins, playa wetlands, aquaculture wetlands, freshwater river deltas, tidal mudflats, coastal marshes, estuaries, and bays. Both migrating and wintering Canvasbacks are particularly fond of wetlands supporting wildcelery, chufa, banana waterlily, and sago pondweed. In San Patricio County oxbow lakes, they used open water areas 45–84 inches deep.

REPRODUCTION

Pair Bonds: Canvasbacks form pair bonds during spring migration or shortly after their arrival at breeding grounds. Paired males follow their mates to breeding areas. Adult females exhibit a high degree of fidelity to breeding areas, returning at rates approaching 75 percent. Pair bonds dissolve in early incubation.

Nesting: Females construct nests 1–7 days prior to egg laying, although construction and maintenance activities may continue throughout incubation. Vegetation used to construct floating platforms is found at the nest site. Nest bowls are lined with small pieces of herbaceous vegetation and down. Only one egg is laid per day, and clutch size is about eight eggs. Only females incubate. Incubating females spend about 92 percent of their time on the nest. Their incubation period is 22–27 days.

In Manitoba their nests were parasitized by other Canvasbacks at rates approaching 36 percent. They occasionally parasitize nests of other species. In Saskatchewan about 67 percent of Canvasback nests were parasitized by Redheads.

Ducklings: Females brood ducklings in the nest until they are dry. They move their young frequently and will feign injury to lead predators away from ducklings. Ducklings are capable of flight at 56–68 days.

APPEARANCE

Breeding: Adult male Canvasbacks are often described as regal. Their overall head and neck coloration is chestnut red, although it grades to blackish brown near their bill and crown. Their breast is black. Their back and sides appear white from a distance, but in hand they are grayer due to fine blackish vermiculation on the white feathers. Their rump and tail are blackish brown, and their tail feathers are short and stubby. Their belly is white. Adult females have a buffy head and neck that grades into a brownish chest. Their upper back is brownish, and their lower back, sides, and flanks are grayish brown. Their belly is white.

Males have a black bill, and females have a dark bill. Both sexes have wedge-shaped bills, and their overall head profile is wedge-shaped. At Catahoula Lake, Louisiana, adult males and females averaged 2.7 and 2.9 pounds, respectively. *Nonbreeding:* Males appear dull in their nonbreeding plumage. The nonbreeding plumage of adult females is similar in appearance to their breeding plumage.

SOURCES

INTRODUCTION: Bluhm 1985; Lovvorn 1990. TEXAS DISTRIBUTION: Traweek 1978; Pulich 1988; Eslinger and Wilson 2001; Seyffert 2001; White 2002; Ray et al. 2003; Lockwood and Freeman 2004; Baar et al. 2008; USFWS 2008. TEXAS HARVEST: Kruse 2007. POPULATION STATUS: NAWMP 2004; USFWS 2011. DIET: Cely 1979; Perry and Uhler 1988; Tome and Wrubleski 1988; Austin et al. 1990; Hohman and Rave 1990; Hohman et al. 1990. RANGE AND HABITATS: Hochbaum 1944; White and James 1978; Bellrose 1980; Stoudt 1982; Serie et al. 1983; Kantrud 1986; Prellwitz 1987; Hohman et al. 1990; Arnold et al. 1995; Maxson and Riggs 1996; Havera 1999b; Mowbray 2002; Baar et al. 2008. REPRODUCTION: Hochbaum 1944; Weller 1959; Sugden 1980; Lovvorn 1990; Austin and Serie 1991; Batt et al. 1992; Sorenson 1993, 1998; Leonard et al. 1996; Anderson et al. 1997; Mowbray 2002. APPEARANCE: Johnsgard 1975; Bellrose 1980; Barzen and Serie 1990; Hohman 1993; Mowbray 2002.

REDHEAD

Aythya americana

No areas are more important to wintering Redheads than the Laguna Madre of Texas and the Laguna Madre de Tamaulipas. From 1955 to 1994 the estimated number of Redheads wintering in the Laguna Madre system averaged 546,000 annually. They have a wintering distribution that includes most of the United States and Mexico, but up to 80 percent of their continental population may winter in the Laguna Madre system. The importance of the Laguna Madre cannot be overstated.

TEXAS DISTRIBUTION

Breeding: Redheads are the most common diving duck in the High Plains during May and June. Broods have been reported in Lubbock and Castro County. In the 1970s, 656 adults were observed during a four-year study in 12 High Plains counties. However, production was low, considering brood sightings averaged less than two per year. In a recent High Plains study, no Redhead ducklings were observed, although adults were present. Broods have also been observed in El Paso, Hudspeth, and possibly Medina Counties.

Redhead (male and female). *Photograph by Greg Lasley, February 12, 2006, Aransas Pass, Texas.*

Migration: Redheads occur throughout Texas during migration. Migrants arrive in the High Plains in September and increase through October. In northern portions of the Rolling Plains, Post Oak Savannah–Blackland Prairies, and Pineywoods, the first migrants arrive in September, and they peak in November. In the Coastal Prairies fall migrants arrive in October and rapidly increase in numbers. Most Redheads wintering in the Coastal Prairies depart during March. Numbers in the High Plains decline drastically during April.
Winter: From 2000 to 2008, Redheads averaged 271,764 during the Texas Mid-winter Waterfowl Survey. However, this midwinter survey does not adequately cover bays, lagoons, and nearshore waters, so a special Redhead survey is conducted in these areas. From 2000 to 2003 and 2005 to 2008, the Redhead survey averaged 617,194 annually. The Texas Mid-winter Waterfowl Survey and the Redhead survey overlap in coverage, so the two estimates cannot be combined. Although most Redheads are found along the coast, they winter throughout the state.

TEXAS HARVEST
From 1999 to 2006, Redhead harvest in Texas averaged 38,444 annually. This was about 13 percent of their annual US harvest.

LONGEVITY

The longevity record for a wild Redhead is 20 years, seven months.

POPULATION STATUS

In 2011, estimated Redhead abundance was 1.4 million. This was the highest estimate since the Waterfowl Breeding Population and Habitat Survey was initiated. The North American Waterfowl Management Plan's population objective is 640,000.

DIET

Redheads forage at the surface and by diving. In North Dakota, prelaying and laying females consumed greater than 99 percent invertebrates. Redheads stopping on the Great Lakes during migration foraged heavily on both pondweed and zebra mussels. In the Laguna Madre of Texas, shoalgrass and mollusks accounted for 79 percent and 13 percent of their winter diet, respectively.

RANGE AND HABITATS

Breeding: Redheads primarily breed in the Prairie Pothole Region and in the valleys of western mountain ranges. They also breed in widely scattered, localized areas throughout North America, including Alaska, south central Mexico, and California's Central Valley. Breeding Redheads use many wetland types, including seasonal wetlands, semipermanent wetlands, and large marshes. They typically nest overwater in dense emergent vegetation, such as cattails. They occasionally nest in uplands and in horizontal nest cylinders (a type of artificial nest structure).
Migration and Winter: Migrants use wetlands throughout North America. In northeastern Texas migrants often use large impoundments. Redheads winter in southwestern Canada, most of the United States, Mexico, and Guatemala. The bulk of their wintering population is found in Apalachee Bay (Florida), Chandeleur Sound (Louisiana), and the Laguna Madre (Texas and Tamaulipas). Freshwater ponds located near the Texas coastline are important to sustaining large Redhead concentrations in the Laguna Madre, as individuals foraging in hypersaline waters require dietary freshwater. Wintering Redheads use a variety of habitats, including large reservoirs and lakes, playa wetlands, freshwater river deltas, coastal marshes, estuaries, and bays.

REPRODUCTION

Pair Bonds: Pair bond formation occurs during winter and spring migration. Along the Texas coast about 35 percent of females were paired in December and January, and more than 58 percent were paired in February. Females select their mates, which follow them to breeding areas. Females frequently return to familiar nesting areas. Pair bonds dissolve during egg laying or incubation.
Nesting: Nest construction begins several days before egg laying. Females construct floating platforms and line nest bowls with small pieces of herbaceous vegetation and down. Construction and maintenance of the platform continues throughout egg laying and incubation. Eggs are laid at a rate of one per day. Only females incubate.

In studies that exclude parasitic eggs, estimated clutch size was seven or eight eggs. Females take multiple incubation breaks per day. Their incubation period is 23 or 24 days.

In Manitoba approximately 36 percent of monitored Redheads laid only parasitic eggs; this group did not construct their own nest or incubate their own eggs. An additional 27 percent of monitored females both laid parasitic eggs and attempted to nest. In Saskatchewan 61 percent of Redhead nests were parasitized by other Redheads. In some locations 60 percent of all Redhead eggs may be laid parasitically. Some authorities suggest 50 percent of all Redhead ducklings may be hatched from parasitically laid eggs. Redheads have been documented to parasitize nests of many other species, including Canvasbacks.

Ducklings: Females brood ducklings in the nest until they are dry and may feign injury to lead predators away from ducklings. Females often abandon the young before they fledge. Ducklings fledge when they are 49–73 days old.

APPEARANCE

Breeding: Adult males have a head and upper neck that are chestnut red. Their lower neck and breast are black. Their back and sides are gray. Their rump and tail are blackish, and their belly is white. Adult females have a yellowish-brown head and neck, but the area behind their bill is typically paler. Females often have scattered white feathers on the back of their head, which may sometimes be dense enough to give their head a frosted appearance. Their breast is brown, and their back, sides, and tail are grayish brown. Their belly is white.

Males have a pale bill with a black tip, and a faint white ring separates the black tip from the pale blue upper bill. Females have a slaty bill with a dark tip, and a pale bluish ring separates the tip from the upper bill. Adult males and females along the Texas coast averaged 2.5 and 2.2 pounds, respectively.

Nonbreeding: Relative to their breeding plumage, males have a neck and head that are browner, and they may have patches of brown plumage on their back and sides. The nonbreeding plumage of adult females is similar to their breeding plumage.

SOURCES

INTRODUCTION: Weller 1964; Bellrose 1980; Michot 2000; USFWS 2011. TEXAS DISTRIBUTION: Oberholser 1974; Traweek 1978; Rhodes 1979; Bellrose 1980; Pulich 1988; Adair et al. 1996; Woodin 1996; Michot 2000; Esslinger and Wilson 2001; Seyffert 2001; White 2002; Ray et al. 2003; Lockwood and Freeman 2004; Baar et al. 2008; USFWS 2008; Johnson et al. 2010. TEXAS HARVEST: Kruse 2007. LONGEVITY: Lutmerding and Love 2011. POPULATION STATUS: NAWMP 2004; USFWS 2011. DIET: Woodin and Swanson 1989; Custer and Custer 1996; Michot et al. 2008. RANGE AND HABITATS: Hochbaum 1944; Low 1945; Weller 1959; Lokemoen 1966; McKnight 1974; Michot et al. 1979; Bellrose 1980; Pulich 1988; Sorenson 1991; Yerkes 1998, 2000; Yerkes and Kowalchuck 1999; Woodin and Michot 2002; Baar et al. 2008. REPRODUCTION: Hochbaum 1944; Low 1945; Weller 1959, 1965; Rienecker and Anderson 1960; Smart 1965; Lokemoen 1966; Joyner 1973, 1976; Sugden and Butler 1980; Sorenson 1991, 1998; Batt et al. 1992; Fleskes 1992; Yerkes 1998, 2000; Woodin and Michot 2002. APPEARANCE: Weller 1957; Johnsgard 1975; Bellrose 1980; Woodin and Michot 2002.

RING-NECKED DUCK

Aythya collaris

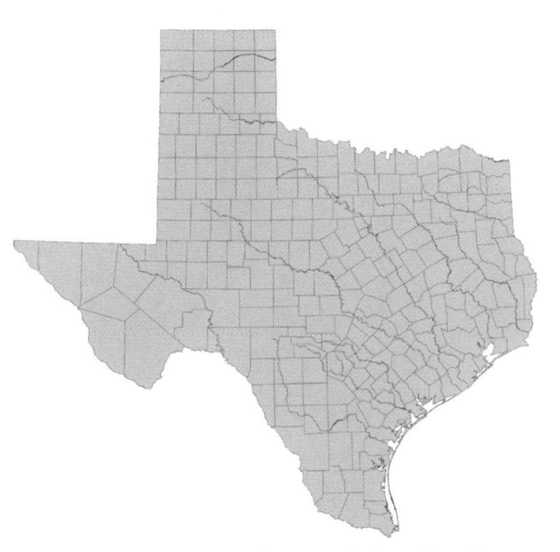

Kleptoparasitism occurs when one bird steals food from another bird. Among waterbirds, kleptoparasitism is most common among individuals that forage on large items that take time to consume, such as fish or long pieces of aquatic vegetation. Ring-necked Ducks commonly forage on aquatic vegetation, which they regularly lose to aggressive American Coots. American Coots take food directly from the bill of the Ring-necked Ducks or chase them until they drop it.

TEXAS DISTRIBUTION

Breeding: Ring-necked Ducks do not nest in Texas, although they are occasionally observed in the western half of the state during summer.

Ring-necked Duck (male). *Photograph by Trey Barron, February 15, 2009, Amarillo, Potter County, Texas.*

Ring-necked Duck (female). *Photograph by Trey Barron, February 15, 2009, Amarillo, Potter County, Texas.*

Migration: Ring-necked Ducks occur throughout Texas during migration. They arrive in the High Plains in early October. In northern portions of the Rolling Plains, Post Oak Savannah–Blackland Prairies, and Pineywoods, the first migrants arrive in early or mid-October, and their numbers peak in November. Fall migrants arrive on the Coastal Prairies in October and steadily increase in number through December and January. Numbers in the Coastal Prairies typically decline in February. In northern portions of the Rolling Plains, Post Oak Savannah–Blackland Prairies, and Pineywoods, spring migrants are most abundant in February or early March, and they may linger into May. Most migrants are gone from the High Plains by April.
Winter: From 2000 to 2008, Ring-necked Ducks averaged 135,638 during the Texas Mid-winter Waterfowl Survey. They are most common in the Rolling Plains, Post Oak Savannah–Blackland Prairies, and South Texas Brush Country (TPWD unpublished). In a survey of flood prevention lakes in the Blackland Prairies, 14 percent of the ducks observed were Ring-necked Ducks.

TEXAS HARVEST

From 1999 to 2006, Ring-necked Duck harvest in Texas averaged 50,714 annually. This was about 10 percent of their annual US harvest.

POPULATION STATUS

Their long-term trend is stable to increasing. From 1994 to 2003 their average population size was about 2 million.

DIET

Ring-necked Ducks primarily forage by diving. In northwestern Minnesota the diet of laying females was 92 percent invertebrates. During fall, wildcelery accounts for over 50 percent of their diet at Lake Onalaska, Wisconsin. Vegetation dominated their diet at Catahoula Lake, Louisiana; the most common foods consumed were chufa flatsedge tubers, barnyard grass seeds, common arrowhead tubers, and bearded sprangletop seeds.

RANGE AND HABITATS

Breeding: Ring-necked Ducks breed from the Island of Newfoundland south to Maine, west through the Prairie Pothole Region, and northwest into Alaska. They also breed locally in the Intermountain West. Breeding pairs settle in on freshwater marshes, large freshwater impoundments, flooded meadows, beaver ponds, and semipermanent and permanent prairie wetlands. They are overwater nesters. Wetlands used for nesting typically have abundant emergent vegetation and are often interspersed with floating aquatic vegetation.
Migration: Migrants use wetlands throughout most of North America. During fall, large concentrations may be found on wetlands that support wild rice, particularly in Minnesota. Habitats used include rainwater basins, playa wetlands, stock ponds, shallow reservoirs, marshes, and beaver ponds.
Winter: Ring-necked Ducks winter in southwestern British Columbia, in the western and southern portions of the United States, and along the East Coast from Massachusetts to Florida. They also winter in the Caribbean, Mexico, and Central America. They are particularly fond of wetlands that support floating plants or submersed aquatic vegetation (for example, hydrilla). They utilize reservoirs, lakes, stock ponds, aquaculture ponds, playa wetlands, beaver ponds, rivers, moist-soil management wetlands, municipal wetlands, and coastal wetlands that are fresh to intermediate in salinity. They are uncommon in brackish and saltwater habitats unless forced into them by freezes. In San Patricio County oxbow lakes, they used water 35–45 inches deep.

REPRODUCTION

Pair Bonds: Most Ring-necked Ducks form pair bonds during spring migration. Females exhibit little fidelity to breeding areas. Pair bonds dissolve in early incubation.
Nesting: Only females construct nests, which are floating platforms, and incubate eggs. Females line nest bowls with down. One egg is laid per day. Clutches typically consist of eight or nine eggs. Their incubation period is about 26 days.
Ducklings: Females brood young in the nest, perhaps up to 4 days, and are vigilant for predators. Females typically remain with young until they fledge, which occurs when ducklings are 49–56 days old.

APPEARANCE

Breeding: Adult males have a glossy, highly iridescent black head, neck, breast, and back. Their head has an angular shape, and both the head and the neck often appear purple. The name Ring-necked Duck is due to a narrow chestnut ring around the base of their neck. Their belly is white and their sides are gray; vertical white lines extend up from their belly and separate the breast and sides. Their rump and tail are blackish. Adult females are brownish with a dark brown back. Their crown is dark brown. The sides of their head are light brown or gray, becoming paler toward the bill. They have light eye rings. Their belly is white.

Males have a bluish bill with a black tip. A broad white line separates the black tip from the bluish upper bill, and a narrow white line outlines the base of their bill. Females have a bluish bill with a faint white ring near the tip. In Minnesota, adult males and females averaged 1.8 and 1.6 pounds, respectively, during fall.

Ring-necked Ducks are often confused with Lesser Scaup. Ring-necked Ducks have dark wings with a gray trailing edge (although, in hand, thin white edging is visible on the tips of their secondaries). Lesser Scaup, in contrast, have a conspicuous broad white stripe along the trailing edge of their wings.

Nonbreeding: In males, this plumage is brownish but may be considerably blotched with black, particularly on the breast and back. The nonbreeding plumage of adult females is similar to their breeding plumage.

SOURCES

INTRODUCTION: Bergan and Smith 1986; LeSchack and Hepp 1995. TEXAS DISTRIBUTION: Hobaugh and Teer 1981; Pulich 1988; Esslinger and Wilson 2001; Seyffert 2001; White 2002; Lockwood and Freeman 2004; Baar et al. 2008. TEXAS HARVEST: Kruse 2007. POPULATION STATUS: Hohman and Eberhardt 1998; NAWMP 2004. DIET: Hohman 1985; Weller 1988; Peters and Afton 1993; Eberhardt and Riggs 1995; Hohman and Eberhardt 1998. RANGE AND HABITATS: White and James 1978; Bellrose 1980; Hohman 1985; Weller 1988; Bergan and Smith 1989; Maxson and Riggs 1996; Hohman and Eberhardt 1998; Koons and Rotella 2003, Baar et al. 2008. REPRODUCTION: Hochbaum 1944; Weller 1959, 1965; Bellrose 1980; Hohman 1986; McAuley and Longcore 1989; Hier 1989; Batt et al. 1992; Maxson and Pace 1992; Maxson and Riggs 1996; Hohman and Eberhardt 1998; Brua 2001. APPEARANCE: Johnsgard 1975; Bellrose 1980; Hohman and Eberhardt 1998.

GREATER SCAUP

Aythya marila

Traditionally, migrating and wintering Greater Scaup in the Great Lakes foraged heavily on gastropods, which accounted for about 90 percent of their diet. However, zebra mussels were introduced to the Great Lakes during the mid- to late 1980s. Within 15 years after their introduction, zebra mussels accounted for about 60 percent of the Greater Scaup diet. There is concern that Greater Scaup for-

Greater Scaup (male). *Photograph by Mark W. Lockwood, March 12, 2005, Balmorhea Lake, Reeves County, Texas.*

aging on zebra mussels may have unhealthy toxin levels, as zebra mussels carry heavy loads of selenium and other contaminants.

TEXAS DISTRIBUTION

Breeding: Greater Scaup do not nest in Texas. They are very rare in spring and occasionally linger into early summer.

Migration: Although rare, fall migrants arrive in northern portions of the Rolling Plains, Post Oak Savannah–Blackland Prairies, and Pineywoods about late October to mid-November. Fall migrants arrive on the Coastal Prairies in mid-November and typically depart by mid-March, although a few may linger into June. Migrants are rarely observed in western portions of the state.

Winter: Greater Scaup are rare to locally uncommon in coastal bays and nearshore waters of the central and upper coast. In East Texas they are rare to irregular visitors on most reservoirs, although they are frequently observed in some areas (for example, Lake o' the Pines). There is no estimate of their wintering numbers in Texas.

TEXAS HARVEST

From 1999 to 2006, Greater Scaup harvest in Texas averaged 1,680 annually; this was about 3 percent of their estimated annual harvest in the United States. The ratio of Greater to Lesser Scaup in the Texas harvest is 1 to 18.3. This ratio should be viewed cautiously, as it is likely reflective of the distribution of the two species with respect to hunting pressure and not reflective of their actual abundance.

POPULATION STATUS

Their long-term trend is likely decreasing. Greater and Lesser Scaup are combined in the Waterfowl Breeding Population and Habitat Survey because of their similarity in appearance. In 2008 their combined estimate was 4.3 million. Greater Scaup accounted for about 11 percent of this estimate.

DIET

Greater Scaup forage primarily by diving. Those wintering in Long Island Sound foraged heavily on sea lettuce and marine gastropods, including whelks. In Apalachee Bay, Florida, their winter diet approached 100 percent animal matter and was composed mainly of marine gastropods (82 percent) and mud crabs (14 percent). Important foods of fall and spring migrants on the Great Lakes included zebra mussels, gastropods, wildcelery, and muskgrass.

RANGE AND HABITATS

Breeding: Greater Scaup primarily breed in tundra zones of North America, Europe, and Asia. In North America they breed in the open boreal forests and tundra zones of western Alaska, locally in the Northwest Territories, in coastal wetlands surrounding Hudson Bay, and across northern Quebec and Labrador. Breeding pairs use open areas with abundant, interspersed shallow lakes and ponds. Nests are typically constructed along the banks of wetlands in residual cover, usually within three feet of the wetland's edge. Nests are also found in low-lying grassy meadows, on islands, on floating mats of vegetation, and in gull and tern colonies.

Migration: Migrants use the Great Lakes and their associated marshes. They also use other large lakes, large man-made reservoirs, large rivers, bays, estuaries, river deltas, and offshore areas.

Winter: Greater Scaup winter along the Pacific coast from Alaska to Baja California and along the Atlantic coast from the Island of Newfoundland to Florida. A few winter in the Gulf of Mexico from the Florida Panhandle west to the Rio Grande. Wintering birds are also found in the Great Lakes and, rarely, in inland reservoirs across the United States. The largest numbers are found in Long Island Sound, where 42–49 percent of the North American population winters. Wintering birds primarily use coastal estuaries, bays, and nearshore waters but are occasionally found in small lakes and impoundments near the coast. They have a high degree of fidelity to both wintering areas and migration routes.

REPRODUCTION

Pair Bonds: Pairs are formed in later winter, during spring migration, or shortly after their arrival at breeding grounds. Paired males follow their mates to breeding areas. Females often nest near locations used in previous years. Pair bonds dissolve in early to mid-incubation.

Nesting: Nest construction and incubation are activities of the female. They line their nests with down and plant litter. If water rises in the immediate area of the nest, females may add vegetation to the base to raise its level. Only one egg is laid per day. The clutch size is normally eight or nine eggs. The incubation period is 23–28 days. If initial nests fail, about 50 percent of females may renest. They will not renest after ducklings hatch. They have been reported to parasitize nests of other ducks.

Ducklings: Females brood ducklings in the nest until they are dry and may periodically brood ducklings after leading them away from the nest. The amalgamation of several broods is common.

APPEARANCE

Breeding: Adult males have a black head, neck, breast, rump, and tail. Their head has a round appearance and generally has greenish iridescence. Their back is pale gray and becomes darker toward the rump. Their sides are whitish and their belly is white. The head and neck of adult females typically vary from buff to dark brown but can even be black with a greenish sheen like those of males. They have a white area on the front of their head near the bill. Occasionally, they may have scattered white feathers on the back of their head and neck. Their breast is dark brown, and their back, sides, rump, and tail are brown. Their belly is white.

Bill color in males is pale blue, and in females, slaty. In Long Island Sound adult males and females averaged 2.3 and 2.1 pounds, respectively, during winter.

Nonbreeding: In adult males, the chin and forward portion of their head are light brown, and the remainder of their head and neck is blackish brown. Unlike females, males typically do not have white on the forward portion of their head. The remainder of their plumage is like that of females. The nonbreeding plumage of adult females is similar in appearance to their breeding plumage.

SOURCES

INTRODUCTION: Custer and Custer 1996; Ross et al. 2005; Badzinski and Petrie 2006. TEXAS DISTRIBUTION: Pulich 1988; Seyffert 2001; White 2002; Lockwood and Freeman 2004; Eubanks et al. 2006. TEXAS HARVEST: Kruse 2007. POPULATION STATUS: Bellrose 1980; NAWMP 2004; USFWS 2011. DIET: Stieglitz 1966; Wahle and Barclay 1993; Kessel et al. 2002; Badzinski and Petrie 2006. RANGE AND HABITATS: Boyd 1974; Bellrose 1980; Gooders and Boyer 1986; Reed et al. 1992; Wormington and Leach 1992; Fournier and Hines 2001; Kessel et al. 2002; Badzinski and Petrie 2006; Flint et al. 2006. REPRODUCTION: Weller 1959; Hildén 1964; Weller et al. 1969; Palmer 1976a; Bellrose 1980; Kessel et al. 2002; Flint 2003; Flint et al. 2006. APPEARANCE: Billard and Humphrey 1972; Kessel et al. 2002.

LESSER SCAUP

Aythya affinis

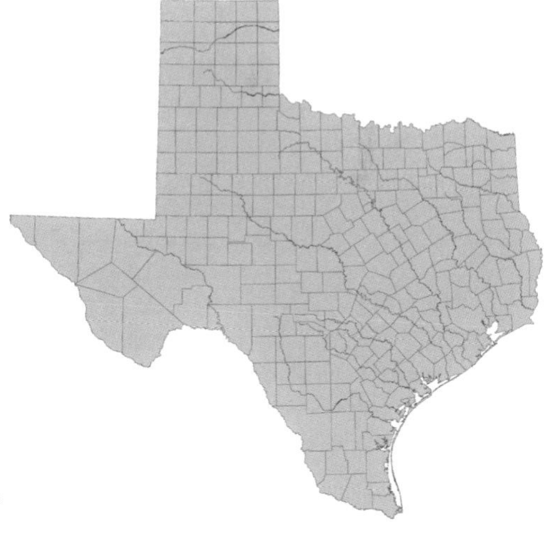

Population numbers of Lesser Scaup have been trending downward since the mid-1980s. Reasons for this trend are unknown, but lipid levels of spring migrants in the upper Midwest (Iowa, Minnesota, and North Dakota) declined during this same period, suggesting habitats they encounter during migration are in poor condition. Additionally, recent studies of Lesser Scaup conducted in the boreal forest have documented low breeding propensity and low nest success, which suggest multiple factors may be contributing to their decline.

Lesser Scaup (male). *Photograph by Trey Barron, February 22, 2008, Amarillo, Potter County, Texas.*

TEXAS DISTRIBUTION

Breeding: Lesser Scaup are rare in summer. There are breeding records from Bailey County (Muleshoe NWR) and Swisher County. The two records are from 1942 and 1977, respectively.

Migration: Lesser Scaup occur throughout Texas during migration. They arrive in the High Plains in late October or early November. They are rather uncommon in the High Plains during winter. In northeast Texas, including portions of the Post Oak Savannah–Blackland Prairies, the first migrants arrive in late October, and they become increasingly common in November. Fall migrants arrive on the Coastal Prairies in October. Numbers in the Coastal Prairies increase greatly during November and remain high through March. They often remain on the Coastal Prairies and in northeast Texas into April and early May. Most are gone from the High Plains by April.

Winter: From 2000 to 2008, Lesser Scaup averaged 136,543 during the Texas Mid-winter Waterfowl Survey. Greater Scaup numbers are included in the Mid-winter Waterfowl Survey estimate, but they likely account for only a small percentage of the birds observed. Lesser Scaup are most common in the Coastal Prairies and South Texas Brush Country but may occur statewide during winter (TPWD unpublished).

TEXAS HARVEST

From 1999 to 2006, harvest in Texas averaged 30,655 annually. This was about 10 percent of their estimated annual US harvest.

LONGEVITY
The longevity record for a wild Lesser Scaup is 18 years, four months.

POPULATION STATUS
Although abundant, their long-term trend is decreasing. In the Waterfowl Breeding Population and Habitat Survey, Lesser and Greater Scaup are combined because of their similarity in appearance. Between 1955 and 2011 their peak population was 8 million. In 2011 their combined population was 4.3 million. About 89 percent of this estimate was Lesser Scaup. The North American Waterfowl Management Plan's population goal for Lesser and Greater Scaup is 6.3 million.

DIET
Lesser Scaup primarily forage by diving and locate food items both tactilely and visually. In Manitoba, breeding Lesser Scaup commonly foraged on midges, caddisflies, and amphipods. In southwestern Louisiana, over 60 percent of the diet of wintering Lesser Scaup was invertebrates, primarily midge larvae. The main foods consumed in the Laguna Madre of Texas were shoalgrass, clams, snails, and crabs.

Lesser Scaup (female). *Photograph by Trey Barron, January 1, 2009, Amarillo, Potter County, Texas.*

Important foods of spring migrants in the Upper Midwest (multiple states and locations) were amphipods, midge larvae, snails, and seeds.

RANGE AND HABITATS

Breeding: Lesser Scaup breed in the boreal forests of Alaska and Canada, the Canadian Parklands, and western portions of the Prairie Pothole Region. A few breed in the western mountain ranges of the United States. Breeding pairs settle in on seasonal and semipermanent wetlands, large lakes, beaver ponds, and shallow reservoirs. Nests are constructed overwater, in uplands, and in moist areas near wetland edges. Upland nests may be located in wet meadows, prairies, shrubby areas, hay fields, and croplands.

Migration: Migrants use wetlands throughout North America. The Great Lakes and their associated marshes are of particular importance, as are impoundments (reservoirs) along the upper Mississippi River. They also use coastal wetlands, reservoirs, playa wetlands, rainwater basins, beaver ponds, rivers, stock ponds, and municipal wetlands.

Winter: Lesser Scaup winter in southwestern British Columbia, along the west and east coasts of the United States, in the Great Lakes, throughout the Mississippi Alluvial Valley, and across the lower third of the United States. The nearshore and offshore waters of the Gulf of Mexico, particularly off the coast of Louisiana, are among their most important wintering areas. They also winter in the Caribbean, Mexico, and Central America. They use reservoirs, lakes, aquaculture ponds, playa wetlands, beaver ponds, municipal wetlands, coastal impoundments, small ponds, coastal marshes, estuaries, and bays. In San Patricio County oxbow lakes, they used areas 35–45 inches deep.

REPRODUCTION

Pair Bonds: Very few (12 percent) Lesser Scaup in Texas were paired in February. Most form pair bonds during spring migration. Once paired, males follow their mates to breeding areas. Many females return to familiar nesting areas. Pair bonds dissolve during incubation.

Nesting: In Alaska, only 68 percent of paired females are estimated to nest on an annual basis. Only females construct nests and incubate eggs. Materials used to build nests are found in the vicinity of nest sites. Overwater nests are typically well-built platforms. Nest bowls are lined with down. Only 1 egg is laid per day, and clutches typically consist of 8–10 eggs. If water rises near the nest, females may add additional vegetation to raise the level of eggs. Their incubation period is about 21–27 days. After failed nesting attempts, some females renest.

Ducklings: Females brood ducklings in the nest until they are dry. They lead their ducklings away from potential predators and will feign injury to lure predators away from the ducklings. They often act aggressively toward other birds that approach their ducklings and have even been observed rushing canoes and boats in attempts to drive them away from their ducklings. The amalgamation of several broods is common, and amalgamated broods may be attended by several females. Females often abandon their young before they fledge; ducklings are capable of flight at 47–50 days.

APPEARANCE

Breeding: Adult male Lesser Scaup have a glossy black head, neck, breast, rump, and tail. Their head peaks slightly toward the back of the crown. Their sides are whitish with varying amounts of gray. Their back is pale gray and grades to dark gray near the rump. Their belly is white. Adult females are dark brown in appearance except for a white area on the front of their head, around the bill. Their belly is white. Eye color of females is assumed to change from brown to yellow as they age. However, eye color cannot be used reliably to age females in the field.

Males have a pale blue bill, and females have a bluish bill. Fall weights in Minnesota averaged 2.1 and 1.9 pounds for adult males and adult females, respectively.

Lesser Scaup are often confused with Ring-necked Ducks. Females of the two species, in particular, are similar in size and appearance. However, Lesser Scaup and Ring-necked Ducks can be separated by their wings. Lesser Scaup have a broad white band along the trailing edge of their wings, and Ring-necked Ducks have gray on the lower portion of their wings.

Nonbreeding: The plumage of males is muted with little sheen. Their head and neck range from drab brown to dull black, and some may have white on the forward portion of their head. Their sides are brownish, and they may have dark mottling on their belly and near their tail. The nonbreeding plumage of adult females is similar in appearance to their breeding plumage.

SOURCES

INTRODUCTION: Walker et al. 2005; Austin et al. 2006; Corcoran et al. 2007; Anteau and Afton 2009; Martin et al. 2009. TEXAS DISTRIBUTION: Hawkins 1945; White and James 1978; Esslinger and Wilson 2001; Seyffert 2001; White 2002; Lockwood and Freeman 2004; Eubanks et al. 2006; Baar et al. 2008; USFWS 2008. TEXAS HARVEST: Kruse 2007. LONGEVITY: Lutmerding and Love 2011. POPULATION STATUS: Bellrose 1980; NAWMP 2004; USFWS 2011. DIET: McMahan 1970; Afton and Hier 1981; Tome and Wrubleski 1988; Afton et al. 1991; Badzinski and Petrie 2006; Anteau and Afton 2008. RANGE AND HABITATS: Keith 1961; Hines 1977; White and James 1978; Bellrose 1980; Weller 1988; Dubovsky and Kaminski 1992; Wormington and Leach 1992; Austin et al. 1998; Fournier and Hines 2001; Kinney 2004; Walker et al. 2005; Badzinski and Petrie 2006; Corcoran et al. 2007; Baar et al. 2008. REPRODUCTION: Hochbaum 1944; Weller 1959, 1965; Vermeer 1968; Hines 1977; Bellrose 1980; Afton 1984, 1993; Lokemoen 1991; Batt et al. 1992; Austin et al. 1998; Koons and Rotella 2003; Walker et al. 2005; Herring and Collazo 2009; Martin et al. 2009. APPEARANCE: Trauger 1974; Austin et al. 1998; Havera 1999a; Vest et al. 2006; Fast et al. 2008.

COMMON EIDER

Somateria mollissima

There is one well-documented record of a Common Eider from Texas. It was a male taken by a hunter in the northern Laguna Madre, Nueces County, on January 8, 2007. This individual was from the population that breeds in northeastern North America, *S. m. dresseri.* This is the most likely subspecies to occur in Texas, as they are regular winter visitors along the southern Atlantic coast.

Common Eiders breed in North America, Europe, and Asia. In North America they breed along the coasts of Alaska, Canada, southern Greenland, and the north-

eastern United States. From 1994 to 2003 the North American population averaged just over 1 million annually. Their numbers are likely declining.

Common Eiders generally nest on islands and narrow peninsulas. Nests are found on the ground in dense coniferous forests, dense shrubs, heath tundra, herbaceous vegetation, and exposed mossy areas. Nests may be isolated or in colonies; larger colonies may involve several thousand females, and nest densities can exceed 100 per acre. Their wintering and breeding ranges are similar, although they tend to occur farther offshore during winter. Their diet includes echinoderms, crustaceans, bivalves, gastropods and herring eggs.

The upper parts of adult males are white except for a thick black band on each side of their white crown. Their belly, sides, rump, and tail are black. They have a yellow to greenish bill. Females vary in color from rusty brown to brown. They are the second largest duck in North America; some Muscovy Ducks may be larger.

Common Eider (adult male). *Photograph by Ron Lockwood, December 3, 2011, in Provincetown, Mass.*

King Eider (first-winter male). *Photograph by Steve Bentsen, May 6, 1998, Quintana, Brazoria County, Texas.*

SOURCES

Cottam 1939; Cantin et al. 1974; Bellrose 1980; Schmutz et al. 1983; Gooders and Boyer 1986; Goudie and Ankney 1986; Cornish and Dickson 1997; Goudie et al. 2000; NAWMP 2004; Lockwood 2008.

KING EIDER

Somateria spectabilis

There are two well-documented King Eider records from Texas, both of which involved first-winter males. They occurred in Brazoria and Galveston Counties in 1998 and 2005, respectively. The first bird was captured and sent to a wildlife rehabilitator. There is also a potential record from Aransas County in 1968. These birds follow the occurrence pattern of other individuals found along the Gulf coast.

King Eiders breed in the extreme arctic regions of North America, Europe, and Asia. In North America they averaged 575,000 annually from 1994 to 2003. Their numbers are declining.

King Eiders nest on the ground in wetland margins, meadows, shrubby areas, and barren hillsides. Their nests are often concealed by rocks or hummocks. They winter along the southern coast of Alaska, in southern Greenland, and along the east coast of Canada south to Maryland. Their diet includes echinoderms, crustaceans, bivalves, gastropods, insects, and algae.

Adult males are spectacular in appearance. Their bill is red with either a pink or white tip. Their frontal lobe, or the area above the base of the bill, is extremely enlarged and bright yellow. The area under their eye is light green, and the remainder of their head is bluish gray. Their neck and upper back are white, and their back, sides, belly, and tail are black. Their breast is pinkish. Females are brown.

SOURCES

Cottam 1939; Gooders and Boyer 1986; Suydam 2000; Lockwood and Freeman 2004.

HARLEQUIN DUCK

Histrionicus histrionicus

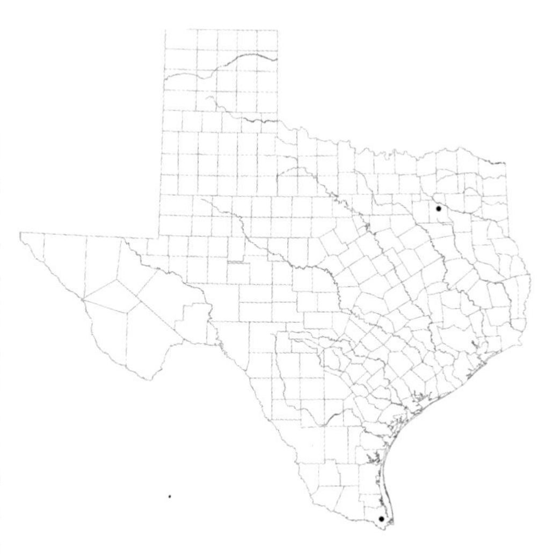

There are two well-documented sightings of Harlequin Ducks from Texas. A male was photographed in Cameron County in 1990, and a male and female (possibly a pair) were observed in Van Zandt County in 1995. There are also three potential records from Aransas County, but these were not as well documented. One of the Aransas County sightings involved a male and female in Copano Bay; a written description of this observation was published in the *Auk* in 1945.

Harlequin Ducks breed in Iceland, eastern Russia, Japan, and North America. There are two breeding populations in North America. One breeds in the Northwest, from Alaska south into Oregon, Idaho, and Wyoming. This population winters along the Pacific coast from the Aleutian Islands south to Oregon. The other population breeds in northeastern Canada and southern Greenland. This population winters off the coast of southern Greenland and off the coast of Newfoundland and Labrador south to Maryland. Their diet includes aquatic insects, gastropods, fish roe, small fish, echinoderms, small crabs, and other crustaceans.

They are North America's only river duck. They breed along swift, fast-flowing

Harlequin Duck (male and female). *Photograph courtesy of Arthur Morris/VIREO, taken in 2000, Ninilichik, Alaska.*

streams. Their nesting habits are extremely variable; nests may be located on the ground, in tree cavities, in rock crevices, or on cliff ledges. Although Harlequin Ducks form seasonal pair bonds (pair bonds are formed during winter and dissolve during the breeding season), they annually re-pair with the same mate as long as the mate is living.

Estimates suggest over 1,000 Harlequin Ducks were directly killed by the *Exxon Valdez* oil spill. Moreover, it took approximately one decade for survival estimates of females wintering in areas impacted by the oil spill to recover to pre-spill levels. From 1994 to 2003 their annual population estimate was roughly 254,000.

Adult males average 1.4 pounds, and they are uniquely colored. Their overall body color is slaty blue. They have a large white crescent between the bill and eye, a small round white patch behind the eye, and a short diagonal white slash on the back of their head and upper neck. They also have a white neck ring and vertical white bar on each side of the breast; both the neck ring and vertical bar are bordered by black. Their sides are cinnamon red, and they have a small cinnamon marking under the upper portion of their facial crescent. Their back is blue and white. Females are brown, with white on the forward portion of their head and behind their eye.

SOURCES

Cottam 1939; Hagar 1945; Gooders and Boyer 1986; Robertson and Goudie 1999; Smith et al. 2000; Esler et al. 2002; Lockwood and Freeman 2004; NAWMP 2004; Esler and Iverson 2010.

SURF SCOTER

Melanitta perspicillata

Surf Scoters are rare in Texas, but they may occur throughout the state between late October and late April. They are most common on the upper and central portions of the Coastal Prairies, where they occur in very low numbers during most winters. When sighted, they are often with Lesser or Greater Scaup. Most Surf Scoters observed in Texas are subadults.

Surf Scoters breed in the boreal forests of Alaska and Canada; however, breeding is localized rather than widespread. There are roughly 600,000 Surf Scoters in North America. Their population is healthy but perhaps declining. They do not breed until they are two to three years old, so their population always has a large nonbreeding segment. Their population does not fluctuate greatly from year to year because of the large number of nonbreeders.

They form seasonal pair bonds. Breeding pairs use small lakes, and females nest in adjacent uplands. They also nest on islands. Females construct nests on the ground, generally near water. Nests are often concealed by logs or low-hanging conifer branches.

Surf Scoters winter in estuaries, bays, and nearshore areas of the Pacific coast from Alaska (including the Aleutian Islands) south to Baja California Sur and along the Atlantic coast from southeastern Canada to Florida. Smaller numbers winter in the Gulf of California and along the northern coast of the Gulf of Mexico. Outside of the breeding season, marine bivalves and herring spawn are important foods.

Females of all ages are brown, and their crown is notably darker than their face. Adult females have faint white plumage near the base of their bill and behind their eye; they also have a faint white patch on the back of their head. Their bills are mottled with greenish black and pale gray. First-winter males are blackish brown. Early in winter they have head markings similar to those of females but have a small white patch on the back of their head just below their crown. They become blacker during late winter, and the white patch on the back of their neck becomes larger. Second-winter males have plumage similar to that of adult males, except that adult males have a large white area on their forehead, and their black plumage has a glossy

Surf Scoter (female). *Photograph by Greg Lasley, December 9, 2009, Austin, Travis County, Texas.*

blue sheen. Their bill is multicolored and large; the front portion is edged in yellow and rapidly grades to orange and then red as it approaches the nostrils. The top one-third of their bill is black. The sides of the bill are pale blue to white, and there is a round black dot near the base. Males weigh about 2.3 pounds. Females are slightly smaller.

SOURCES

Johnsgard 1975; Bellrose 1980; Vermeer 1981; Savard et al. 1998; Seyffert 2001; Sullivan et al. 2002; White 2002; Lockwood and Freeman 2004; NAWMP 2004; Eubanks et al. 2006; Anderson et al. 2008.

WHITE-WINGED SCOTER

Melanitta fusca

White-winged Scoters are rare to very rare in Texas. They potentially occur on wetlands throughout the state, but like Surf Scoters, they are most common on the upper and central portions of the Coastal Prairies. Most White-winged Scoters observed in Texas are first-winter birds. They primarily occur between early November and mid-March.

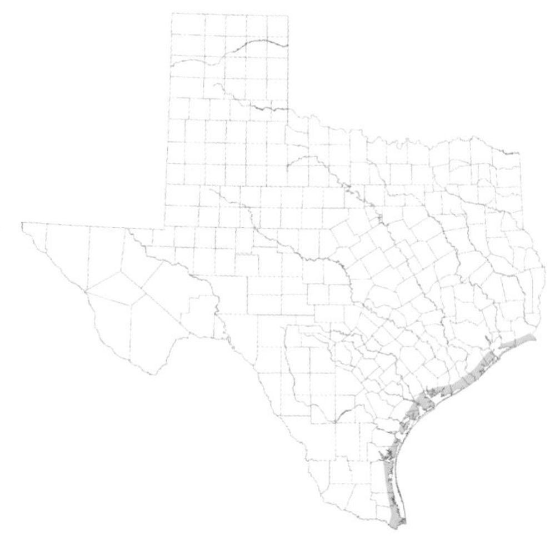

White-winged Scoters breed across northern Europe and Asia and in the boreal forests of Alaska and northwestern Canada. Scattered breeding also takes place in the Canadian Parklands. There are roughly 600,000 White-winged Scoters in North America. Their population trend is likely declining.

White-winged Scoters form seasonal pair bonds. Breeding pairs use large lakes, particularly those with islands. They nest on the ground, often among low-growing shrubs.

They winter in estuaries, bays, and nearshore areas of the Pacific coast from Alaska (including the Aleutian Islands) south to Baja California and along the Atlantic coast from southeastern Canada to Florida. Wintering concentrations off the coast of British Columbia can exceed 10,000 birds. A few winter in the Great Lakes, in the Gulf of California, and along the Gulf of Mexico's northern coast. Outside of the breeding season, they forage on oysters, scallops, blue mussels, and other marine bivalves.

White-winged Scoter (female-type). *Photograph by Greg Lasley, November 17, 2005, Austin, Travis County, Texas.*

Females of all ages are brown, although adult females are darker brown. Females have faint white plumage near the base of their bill and on their cheeks, and the bill of females is grayish black. First-winter males are blackish. Adult males are also black but have a tear-shaped white area around their eye. Their bill is yellowish orange toward the front and sides and has a large black hump near the base. Some authorities consider White-winged Scoters and Velvet Scoters, which are found in Europe, to be the same species. Males and females weigh about 3.8 and 3.2 pounds, respectively.

SOURCES

Cottam 1939; Bellrose 1980; Brown and Brown 1981; Vermeer and Bourne 1984; Gooders and Boyer 1986; Brown and Fredrickson 1997; Lockwood and Freeman 2004; NAWMP 2004; Eubanks et al. 2006; Žydelis et al. 2006; Badzinski et al. 2008; Safine and Lindberg 2008.

BLACK SCOTER

Melanitta americana

Black Scoters are encountered less frequently in Texas than are the other scoters, making them very rare. They are most common between early November and mid-March. They primarily occur in saltwater, particularly the nearshore waters of the upper and central coasts. They are seldom encountered inland. Most Black Scoters observed in Texas are first-winter birds.

Black Scoters breed in eastern Siberia, in arctic and subarctic regions of Alaska, in extreme northwestern Canada, and in northeastern Canada. There are roughly 400,000 Black Scoters in North America. As with the other scoters, their populations are thought to be declining.

They form seasonal pair bonds. Breeding pairs use shallow tundra lakes and small boreal forest lakes. They nest on the ground, often in brush (such as dwarf birch) or on hummocks covered in residual grasses. Nests are typically located near water.

Black Scoters winter in estuaries, bays, and nearshore areas. Wintering Black Scoters are found along the Pacific coast from Alaska south to Baja California and along the Atlantic coast from southeast Canada to Florida. Small numbers winter in the Great Lakes and along the northern coast of the Gulf of Mexico. Outside of the breeding season, they forage heavily on barnacles, blue mussels, and other marine bivalves.

Females of all ages are brown. The upper part of their head is dark brown, and their cheeks and upper neck are light, tawny brown. First-winter males are black, and adult males are completely black and highly iridescent. Adult males have a black

Black Scoter (female-type). *Photograph courtesy of Glenn Bartley/VIREO, taken November 13, 2007, Qualicum Beach, British Columbia.*

bill with a bright yellow, swollen base; the yellow area has a notable ridge. Males and females weigh about 2.5 and 2.2 pounds, respectively.

SOURCES

Cottam 1939; Bellrose 1980; Vermeer and Bourne 1984; Gooders and Boyer 1986; Bordage and Savard 1995; Lockwood and Freeman 2004; NAWMP 2004; Eubanks et al. 2006.

LONG-TAILED DUCK

Clangula hyemalis

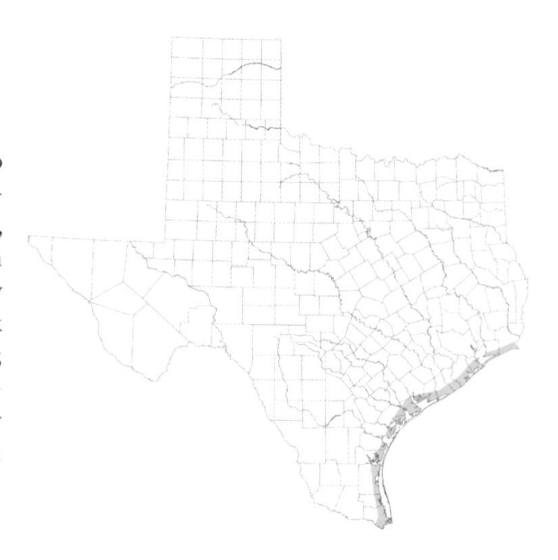

Long-tailed Ducks are rare to very rare in Texas. They potentially occur throughout the state, although they are most common in the Coastal Prairies. Only one banded Long-tailed Duck has been recovered in Texas; surprisingly, the recovery occurred in the High Plains. They tend to occur in Texas between mid-November and early April.

This species was formerly known as Oldsquaw. They have a circumpolar distribution, breeding across the arctic and subarctic regions of Europe, Asia, and North America. Breeding pairs use both freshwater streams and ponds. They also use offshore islands that have freshwater and suitable nesting cover. They nest on the ground and often very close to one another in small colonies. Their pair bonds are seasonal.

Long-tailed Ducks primarily winter along the northern Pacific and Atlantic coasts. They are also found in Hudson Bay, southern Greenland, and the Great Lakes. Wintering birds mainly use estuaries, bays, and nearshore areas. Outside of the breeding season, their diet includes gastropods, bivalves, and aquatic worms.

From 1994 to 2003 the North American population was estimated to number 1 million annually. Their population has declined by almost 50 percent since the late 1970s.

Male and female Long-tailed Ducks weigh about 2.4 and 1.6 pounds, respectively. They are the only North American duck with three, instead of two, plumages per year. In fact, they have one of the most complex plumages of any bird. Most of their plumages consist of white, gray, black, and brown in varying amounts. In all plumages males are notable for their long pintail-like black tail. The plumage worn during winter largely corresponds to the breeding plumage of most other ducks. In this plumage, males have a white head, neck, and upper breast. The sides of their head, however, are gray, and the sides of their upper neck are black. The lower portion of their breast is black, and their back is covered by long gray feathers. Their belly is

Long-tailed Duck (female). *Photograph by Lawrence Semo, January 22, 2010, Austin, Travis County, Texas.*

white. Their bill is black with a wide orange band across the middle. Females are brown during winter but have a black crown and white face. Most males observed in Texas are first-year birds, which are similar in appearance to adults. However, the sides of their head and upper neck are brownish. Other plumages of Long-tailed Ducks are more brown and black than their winter plumage.

SOURCES

Drury 1961; Alison 1975; Peterson and Ellarson 1977; Rofritz 1977; Bellrose 1980; Gooders and Boyer 1986; Goudie and Ankney 1986; Robertson and Savard 2002; Lockwood and Freeman 2004; NAWMP 2004; Eubanks et al. 2006; Schamber et al. 2009; USGS 2010.

BUFFLEHEAD

Bucephala albeola

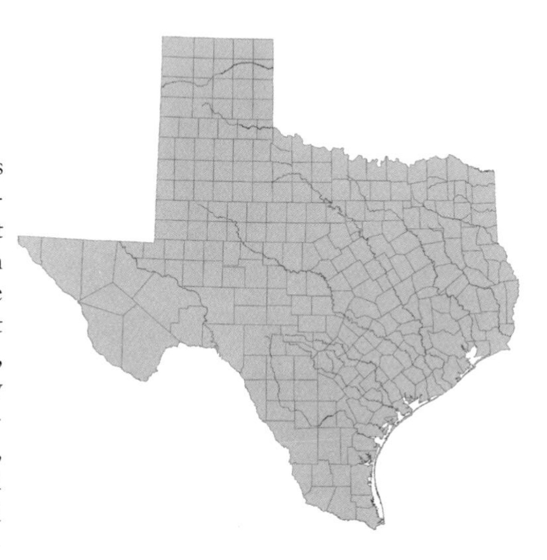

All eight cavity-nesting ducks in North America are secondary cavity nesters, meaning that they do not excavate their own nesting holes but instead use cavities that develop as a result of fire damage, storm damage, or rot, or cavities excavated by woodpeckers. Barrow's Goldeneyes, Common Goldeneyes, Common Mergansers, Hooded Mergansers, Wood Ducks, and Buffleheads regularly use cavities excavated by Pileated Woodpeckers. However, Buffleheads also use cavities excavated by Northern Flickers; in fact, these cavities are their primary nesting sites. Other cavity-nesting ducks are too large to use Northern Flicker holes.

TEXAS DISTRIBUTION

Breeding: Buffleheads do not breed in Texas and only rarely linger into spring and summer.

Migration: Buffleheads are common migrants across much of the state. In the High Plains they are more common during migration than in winter. In northeast Texas, including portions of the Pineywoods and Post Oak Savannah–Blackland Prairies, migrants arrive in late September or early October and remain through late April or early May. They arrive on the Coastal Prairies in early November and remain through late April.

Winter: Wintering Buffleheads potentially occur throughout the state. From 2000 to 2008 they averaged 20,542 during the Texas Mid-winter Waterfowl Survey (TPWD unpublished). Most winter in the Coastal Prairies and Rolling Plains (TPWD unpublished).

Bufflehead (adult male). *Photograph by Raymond S. Matlack, December 17, 2006, South Padre Island, Cameron County, Texas.*

Bufflehead (female). *Photograph by Mark W. Lockwood, January 17, 2009, Midland, Midland County, Texas.*

TEXAS HARVEST

From 1999 to 2006, harvest in Texas averaged about 4,400 annually. This was about 2.6 percent of their annual US harvest. Harvest of Buffleheads is biased toward adult males.

POPULATION STATUS

From 1994 to 2003 their population was estimated to number 1.4 million annually. Their overall population trend is increasing. However, this species is not secure because the boreal forest is under increasing pressure.

DIET

Buffleheads mainly forage by diving. Breeding females in British Columbia consumed over 90 percent animal matter. In Lake Ontario, common foods consumed during winter included midge larvae, amphipods, and snails. In California, foods of spring migrants included water boatman, midge larvae, saltmarsh bulrush seeds, and sago pondweed seeds.

RANGE AND HABITATS

Breeding: Buffleheads breed from the boreal forests of eastern Quebec west-northwest into Alaska. They also breed in the Canadian Parklands, locally in the northern Prairie Pothole Region, and locally in the Intermountain West. Breeding pairs are primarily associated with small lakes and ponds. They favor cavities excavated by Northern Flickers, as they are frequently evicted from larger cavities by Common or Barrow's Goldeneyes. Nest cavities are typically found 2–45 feet above ground and within 100 feet of water. Artificial cavities are readily used.
Migration and Winter: Migrants occur throughout much of North America. Wintering Buffleheads are found along the West Coast from Alaska south to Baja California and along the East Coast from New Brunswick south to Florida. They also winter throughout much of the interior United States and in northern Mexico. The highest wintering densities are found around Vancouver Island, British Columbia. Wintering Buffleheads use protected bays and estuaries, coastal marshes, stock ponds, reservoirs, and slow-moving rivers.

REPRODUCTION

Pair Bonds: Pair bonds are formed during late winter or spring of their second year. Re-pairing with the same mate in successive years has been documented multiple times and may be somewhat common. Males follow their mates to breeding areas. Individual females frequently reuse the same nest cavities. Pair bonds dissolve during incubation.
Nesting: Females select nest sites. They will use bare cavities or those with litter (such as leaves), but they do not carry materials to the nest. They add down during egg laying. Eggs are deposited 1–4 days apart. Clutch size ranges from 6 to 11 eggs. Females incubate for 28–33 days. They do not renest if their clutch is lost after incubation starts. They occasionally parasitize the nests of other cavity-nesting ducks.
Ducklings: Females call ducklings from the nest 24–36 hours after they hatch. Fe-

males brood ducklings, will attempt to drive other ducks away from their young, and will also lead them across lakes or land in search of better foraging areas. The amalgamation of broods is common; amalgamated broods are typically attended by only one female. Females abandon young when they are five to six weeks old. Ducklings fledge at 45–55 days.

APPEARANCE

Breeding: Adult males have a black head, hind neck, back, and rump. They have a large, conspicuous white wedge-shaped patch on their puffy, iridescent black head. Their breast, fore-neck, sides, and belly are white. Adult females have a dark brown head, upper neck, and back, and they have a small, elliptical white patch below their eye. The front of their breast is whitish, and their sides and belly are dark gray.

Adult males have a bluish-gray bill, and adult females have a dark bill. In spring, male and female Buffleheads in California averaged 1.6 and 1.5 pounds, respectively.

Nonbreeding: In both males and females, this plumage is similar to the breeding plumage of adult females. However, males tend to have darker, slightly iridescent heads and larger white head patches.

Subadult: Both immature females and immature males look similar to adult females during their first winter. Males do not acquire their complete black-and-white adult breeding plumage until their second fall.

SOURCES

INTRODUCTION: Gauthier 1993a; Dugger et al. 1994; Eadie et al. 1995, 2000; Hepp and Bellrose 1995; Mallory and Metz 1999. TEXAS DISTRIBUTION: Bolen and Chapman 1981; Seyffert 2001; White 2002; Lockwood and Freeman 2004; Eubanks et al. 2006; Baar et al. 2008. TEXAS HARVEST: USFWS 2005, 2006, 2007a, 2007b. POPULATION STATUS: NAWMP 2004. DIET: Gammonley and Heitmeyer 1990; Thompson and Ankney 2002; Schummer et al. 2008. RANGE AND HABITATS: Erskine 1959, 1960, 1972; Boyd 1974; Bellrose 1980; Limpert 1980; Vermeer 1982; Peterson and Gauthier 1985; Gauthier 1987b, 1987c, 1988, 1989, 1993a; Gauthier and Smith 1987; Gammonley and Heitmeyer 1990; Wormington and Leach 1992; Knutsen and King 2004. REPRODUCTION: Erskine 1972; Bellrose 1980; Savard 1985, 1987; Gauthier 1987a, 1987b, 1989, 1990; Gammonley and Heitmeyer 1990; Evans et al. 2002. APPEARANCE: Palmer 1976b; Bellrose 1980; Gammonley and Heitmeyer 1990; Gauthier 1993a.

COMMON GOLDENEYE

Bucephala clangula

Common Goldeneyes have a circumpolar range, breeding across the boreal forests of North America, Europe, and Asia. They are 1 of 19 Texas waterfowl that breed in both the New and Old Worlds. This group includes Barrow's Goldeneyes, Black Scoters, Brant, Common Eiders, Common Mergansers, Fulvous Whistling-Ducks, Gadwall, Greater Scaup, Greater White-fronted Geese, Green-winged Teal, Harlequin Ducks, King Eiders, Long-tailed Ducks, Mallards, Northern Pintails, Northern Shovelers, Red-breasted Mergansers, and Snow Geese. Two additional North American species, Ruddy Ducks and Canada Geese, have been introduced to Europe.

TEXAS DISTRIBUTION

Breeding: Common Goldeneyes do not breed in Texas.

Migration and Winter: Migrants are rare to uncommon across much of the state. Common Goldeneyes are most common from November through March. They rarely linger into spring. From 2000 to 2008 they averaged 7,142 during the Texas Mid-winter Waterfowl Survey (TPWD unpublished). Most are found in marshes and estuaries of the Coastal Prairies (TPWD unpublished). Their distribution in the Coastal Prairies is skewed; they are locally common along the upper coast and very rare along the lower coast. They are also locally common on some large reservoirs in central and eastern Texas.

Common Goldeneye (adult male). *Photograph courtesy of Glenn Bartley/VIREO, taken January 14, 2008, Victoria, British Columbia.*

Common Goldeneye (female). *Photograph by Mark W. Lockwood, January 9, 2010, Balmorhea Lake, Reeves County, Texas.*

TEXAS HARVEST

From 1999 to 2006, harvest in Texas averaged 850 birds annually. This is about 1 percent of their estimated annual US harvest.

POPULATION STATUS

Common Goldeneyes likely have a stable population. From 1994 to 2003 the North American population was estimated to number 1.3 million annually.

DIET

Common Goldeneyes primarily forage by diving. During the breeding season, caddis fly larvae, dragonfly nymphs, and damselfly nymphs are important foods. Wintering Common Goldeneyes on the Detroit River and in Lake Ontario primarily consumed wildcelery, amphipods, and zebra mussels.

RANGE AND HABITATS

Breeding: Common Goldeneyes breed in the boreal forests of North America from Alaska through eastern Canada. They also breed in the Canadian Parklands, around the Great Lakes, and locally along the US–Canada border. Breeding pairs are primarily associated with rivers and medium- to large-sized lakes. Preformed, existing cavities are required for nesting. They will use natural cavities, rock crevices, and nest boxes.

Migration and Winter: They migrate throughout much of North America. They winter on the west and east coasts of Canada, across most of the United States, and in Mexico. They use rivers, reservoirs, lakes, and protected bays and estuaries.

REPRODUCTION

Pair Bonds: Pair bonds are likely formed during late winter or spring migration. Courtship displays occur on Texas wintering grounds prior to spring migration. Paired males follow their mates north in spring. Females frequently reuse the same nest cavities. Pair bonds dissolve during incubation.
Nesting: Females do not nest until at least their second spring, and up to 64 percent may not nest until their third spring or later. Females select the nest cavity. They lay approximately one egg every other day, and their clutch size is about seven eggs. Down is added during egg laying. Only females incubate. Their incubation period is about 28–32 days. Females are unlikely to renest if their clutch is lost, and they will not renest after ducklings hatch. Common Goldeneyes have been documented to parasitize the nests of other cavity-nesting ducks that share their breeding range.
Ducklings: Females call ducklings from the nest 24–36 hours after they hatch. They brood ducklings and lead them away from potential predators. The amalgamation of multiple broods is common. Females typically abandon their young before they can fly. Ducklings fledge at 56–65 days.

APPEARANCE

Breeding: Both males and females have a triangular-shaped head. Adult males have a black head with a glossy greenish sheen. They have a white oval patch between their eye and bill. Overall, they appear mostly white with black upper parts. Their neck, breast, sides, and belly are white, and their back, rump, and tail are black. The head and upper neck of adult females are chestnut brown, and their eye is yellow. Their lower neck and belly are white. Their breast and body are gray.

Adult males have a black bill. The bill of adult females is usually dark near the base and yellowish orange near the tip. Adult males and females average 2.5 and 1.6 pounds, respectively.
Nonbreeding: In both sexes this plumage is similar to the breeding plumage of adult females. However, males have a dark brown head and neck with a blackish hue, and females have a dull brown head and neck.

SOURCES

INTRODUCTION: Johnsgard 1978; Gooders and Boyer 1986, Mowbray et al. 2000. TEXAS DISTRIBUTION: Pulich 1988; Seyffert 2001; White 2002; Lockwood and Freeman 2004; Eubanks et al. 2006. TEXAS HARVEST: Kruse 2007. POPULATION STATUS: Eadie et al. 1995; NAWMP 2004. DIET: Jones and Drobney 1986; Eadie et al. 1995. RANGE AND HABITATS: Boyd 1974; Bellrose 1980; Lumsden et al. 1980; Afton and Sayler 1982; Reed et al. 1992; Wormington and Leach 1992; Eadie et al. 1995; Savard and Robert 2007. REPRODUCTION: Johnsgard 1978; Bellrose 1980; Afton and Sayler 1982; Eadie et al. 1987, 1988, 1995; Savard 1988; Zicus and Hennes 1988; Zicus 1990; Gauthier 1993a; Mallory and Weatherhead 1993; Wayland and McNicol 1994; Evans et al. 2002; Schmidt et al. 2005; Schummer et al. 2008; Sénéchal et al. 2008. APPEARANCE: Palmer 1976b; Bellrose 1980; Eadie et al. 1995.

Barrow's Goldeneye (adult male). *Photograph courtesy of Rick & Nora Bowers/VIREO, taken in Seward, Alaska.*

BARROW'S GOLDENEYE

Bucephala islandica

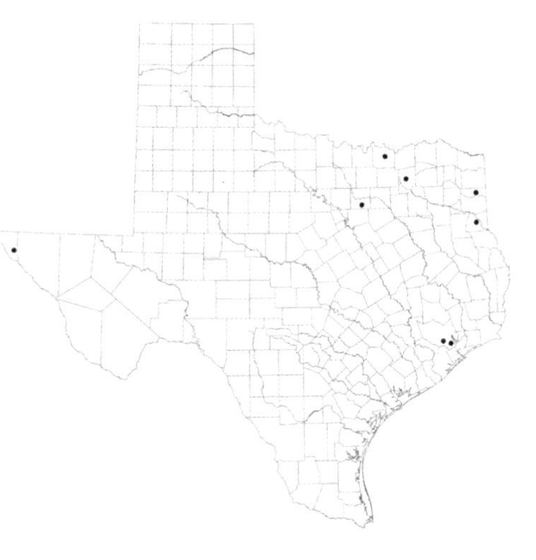

There are nine well-documented Barrow's Goldeneye records from Texas since 1958; all but one occurred east of the 100th meridian, and several involved birds taken by hunters. Most of these records have involved adult males.

Barrow's Goldeneyes breed from Alaska south into the northwestern United States. They also breed locally in Quebec and Iceland. They nest in both natural cavities and nest boxes. They winter in bays and estuaries along the east and west coasts of Canada and the northern United States. Small numbers also winter in the Great Lakes, and a few winter in the western mountain ranges of the United States, including northern New Mexico. Their winter diet is much like that of Common Goldeneyes.

Both sexes look similar to Common Goldeneyes. Adult males have less white overall and have a crescent-shaped white area between the eye and the bill. Common

Goldeneyes, in contrast, have a round to oval white patch in that area. Male Barrow's Goldeneyes also have a row of oblong white markings along their upper sides.

SOURCES

Savard 1985; Eadie et al. 2000; Robert et al. 2000; Lockwood and Freeman 2004.

HOODED MERGANSER

Lophodytes cucullatus

Hooded Mergansers have been an unintended beneficiary of nest boxes built for Wood Ducks. They readily use these boxes, and many local populations have been established as a result of nest box programs. Nest boxes have not only played a role in helping this once rare duck rebound, but they have also helped advance our understanding of this species. Females nesting in boxes can be easily monitored compared to those using natural cavities.

TEXAS DISTRIBUTION

Breeding: Hooded Mergansers are rare, but perhaps increasing, breeders in Texas. Since the late 1970s nesting has been documented in at least eight counties (with evidence for probable breeding in others) spanning the Coastal Prairies, Pineywoods, Post Oak Savannah–Blackland Prairies, and Rolling Plains. Many of these records are related to their use of nest boxes.

Migration and Winter: Migrating and wintering Hooded Mergansers are abundant in Texas and potentially occur on wetlands throughout the state, although they are uncommon on playa wetlands. In northern portions of the Rolling Plains and Post Oak Savannah–Blackland Prairies they are most common between mid-November and early March.

TEXAS HARVEST

From 1999 to 2006, estimated Hooded Merganser harvest in Texas was 5,127 annually. This was about 6 percent of their annual US harvest.

POPULATION STATUS

Hooded Mergansers have a stable, if not increasing, population. Christmas Bird Counts suggest their wintering numbers increased in 39 of 49 states (excluding Hawaii) between 1986 and 2006.

Hooded Merganser (male). *Photograph by Trey Barron, February 15, 2009, Amarillo, Potter County, Texas.*

Hooded Merganser (immature male, with black and white plumage visible on the head). *Photograph by Mark W. Lockwood, April 4, 2008, Balmorhea State Park, Reeves County, Texas.*

Hooded Merganser (female). *Photograph by Trey Barron, December 31, 2011, Amarillo, Potter County, Texas.*

DIET

Hooded Mergansers primarily forage by diving. Common foods include aquatic insects, crayfish, and fish less than three inches in length. They likely hunt by sight, as they consume brook trout that contrast with the environment more frequently than cryptically colored ones.

RANGE AND HABITATS

Breeding: Hooded Mergansers breed in forested regions of the eastern United States and southeastern Canada. They also breed in California, the northwestern United States, and western Canada. Breeding pairs are associated with forested riparian areas and forested wetlands. Preformed, existing cavities are required for nesting. Nest cavities may be over land or water. Nest cavity entrances range from 3 feet to more than 88 feet above ground.

Migration and Winter: They migrate across much of the United States and Canada. They winter in western Canada, southern Ontario, and the United States. Their distribution is localized in the central United States. They use coastal bays, estuaries, tidal creeks, rivers, forested wetlands, reservoirs, shrub-scrub wetlands, and beaver ponds.

REPRODUCTION

Pair Bonds: Hooded Mergansers form seasonal pair bonds. Pairs are observed as early as November. Females likely select new mates each year. Males follow their mates to breeding locations. Females regularly use the same nest cavity from one year to the next. Pair bonds dissolve during incubation.

Nesting: Hooded Mergansers nest in natural or man-made cavities. They do not carry vegetation to the nest, but they do add down. Females lay approximately 1 egg every one to two days until the clutch is complete. Clutch size is typically 10–11 eggs. Only females incubate. They spend about 85 percent of their time on the nest. Their incubation period is 29–33 days. Little is known about their propensity to renest if the first clutch is lost, but they will not renest after a clutch hatches. They occasionally parasitize nests of other cavity-nesting ducks.
Ducklings: All eggs in a clutch hatch within a span of about four hours. Females fly from the nest and give vocal cues to encourage their ducklings to exit. Females will feign injury to lead predators away from ducklings. Young fledge at roughly 70 days.

APPEARANCE

Breeding: Hooded Mergansers exhibit strong sexual dimorphism. Males do not acquire their breeding plumage until their second fall or winter. Adult males have a large white crest that is bordered by black. The crest can be raised and lowered; when raised, it is fanlike, and their head appears large and mostly white. The remainder of their head, neck, and back are black. Long black and white tertial feathers give their lower back a barred appearance. Their rump and tail are brown. Their breast and belly are white. Their breast and sides are separated by a distinctive vertical white stripe that tapers up from the lower breast; this stripe is bordered on each side by black bars that taper down from the upper back. Their sides are tawny or cinnamon and are highly vermiculated. Adult females have a grayish-brown head, neck, breast, and sides. Their crest is bushy, rufous, and small. Their back, rump, and tail are brown. Their belly is white.

The bill of males is black. Females have an upper bill that is black and edged in orange and a lower bill that is yellow. Both sexes have narrow serrated bills; these sawlike serrations have given them and the other mergansers the common name sawbill duck. Adult males and females weigh approximately 1.6 and 1.5 pounds, respectively.
Nonbreeding: Adult males have a reduced crest and a dull, dark brown color; their overall appearance is like that of females. Nonbreeding females have a reduced crest and a browner appearance than breeding-plumaged females.
Immature/subadult: In males, the overall plumage is like that of females. However, the head and neck may have black feathers, and the crest may show some white. In females, this plumage is similar to their breeding plumage.

SOURCES

INTRODUCTION: Dugger et al. 1994. TEXAS DISTRIBUTION: Pulich 1988; Benson and Arnold 2001; Lockwood and Freeman 2004; Davis and Capobianco 2006; Eubanks et al. 2006. TEXAS HARVEST: Kruse 2007. POPULATION STATUS: Dugger et al. 1994; Davis and Capobianco 2006. DIET: Cottam and Uhler 1937; Bellrose 1980; Donnelly and Whoriskey 1991; Dugger et al. 1994; Reimchen 1994. RANGE AND HABITATS: Morse et al. 1969; Bellrose 1980; Doty et al. 1984; Dugger et al. 1994. REPRODUCTION: Morse et al. 1969; Bellrose 1980; Doty et al. 1984; Allen et al. 1990; Zicus 1990; Dugger et al. 1994; Sénéchal et al. 2008. APPEARANCE: Bellrose 1980; Dugger et al. 1994.

COMMON MERGANSER

Mergus merganser

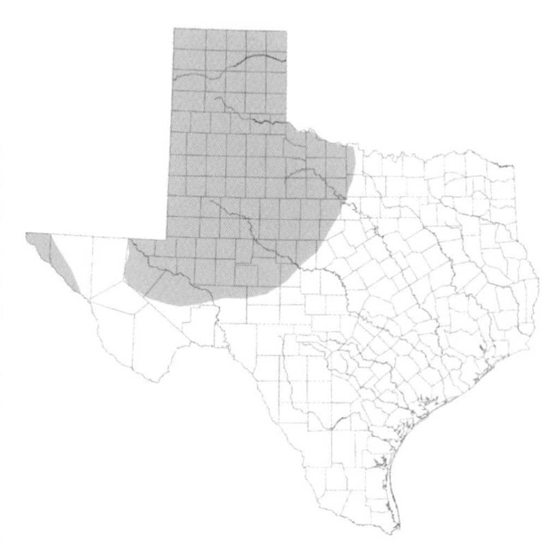

Throughout most of their range, Common Mergansers (known as Goosanders in Europe) are renowned for eating salmon. Because of this, they are often disliked by fishermen, and some countries have even experimented with programs to reduce their numbers. However, they also eat predatory fish that readily consume salmon, and there is little evidence to suggest the reduction in young salmon caused by Common Mergansers is detrimental to populations.

TEXAS DISTRIBUTION

Breeding: Common Mergansers are not known to breed in Texas. There are a handful of summer records, most of which are from northwestern Texas.
Migration and Winter: They are most common in the High Plains and Rolling Plains, where winter concentrations of several hundred to a few thousand sometimes occur on reservoirs. They are locally common in the western Trans-Pecos and rare elsewhere. Fall migrants tend to arrive in mid-November. They begin departing Texas in late March or early April.

TEXAS HARVEST

Common Merganser harvest in Texas is not estimated, but it is likely small.

POPULATION STATUS

Common Mergansers likely have a stable population, but population estimates are poor.

DIET

Common Mergansers dive and hunt for food by sight. They primarily forage on fish, particularly salmon, trout, and shad. Small gizzard shad (< 4.3 inches) and threadfin shad (< 4.4 inches) were the most common fish consumed in Oklahoma and New Mexico, respectively. They occasionally consume crayfish and large aquatic insects.

RANGE AND HABITATS

Breeding: Common Mergansers breed in the boreal forests of North America, Europe, and Asia. In North America they also breed in forested regions of the Intermountain West and locally in western South Dakota. Breeding pairs are associated with large lakes and rivers that have mature forests nearby. They nest in tree cavities,

Common Merganser (male). *Photograph by Mark W. Lockwood, May 11, 2009, Hinman Lake, Colorado.*

rock crevices, nest boxes, and holes in earthen banks (for example, riverbanks).
Migration and Winter: Common Mergansers migrate throughout most of the United States. They winter in coastal regions of southern Alaska, western Canada, and southeastern Canada. They also winter throughout much of the continental United States and Mexico. They are rare to local in the southeastern United States. Wintering Common Mergansers are most common in rivers and reservoirs.

REPRODUCTION
Pair Bonds: Common Mergansers form pair bonds in late February. Males follow their mates north in spring. Pair bonds are seasonal, dissolving after the female begins incubation. Females likely select new mates each year. Pair bonds are not formed until their second winter (or later).
Nesting: Females may carry grasses and other small items to the nest cavity if adequate materials are not already present. Down is added during egg laying. Approximately 1 egg is laid every 1–2 days. Clutch size is 9–12 eggs. Only females incubate. Their incubation period is about 32 days. They frequently parasitize nests of other Common Mergansers.
Ducklings: Ducklings may remain in the nest 1–2 days after they hatch. Females provide vocal cues to encourage ducklings to exit the nest. They brood young at night and during periods of extreme weather. Young occasionally climb onto the backs of females. The period of time that females remain with ducklings is short, perhaps as brief as 7 days. Ducklings fledge at 60–75 days.

APPEARANCE
Breeding: Adult males have a black head and upper neck, both of which have a greenish sheen. Their lower neck, breast, and sides are white. Their back is black, and their

Common Merganser (female). *Photograph by Raymond S. Matlack, December 5, 2007, Canyon, Randall County, Texas.*

rump and tail are gray. Their breast and belly may be light gray or pinkish. Except for a small white chin patch, adult females have a rusty head with a shaggy crest; the rust plumage of their head continues through the middle of the neck. Their lower neck, upper breast, and body are gray. Their lower breast and belly are lighter (almost white).

Males have a red bill, and females have a reddish brown bill. Both sexes have long, narrow serrated bills. Adult males and females average 3.6 and 2.7 pounds, respectively.

Nonbreeding: In both males and females, this plumage is similar to the breeding plumage of adult females. However, males tend to have a shorter crest than that of females, and they may also have an olive-tinged crown.

Immature/subadult: This plumage is acquired their first fall or winter. In males, the overall plumage is like that of females, but their head and neck may have black feathers. In females, this plumage is similar to their adult plumage.

SOURCES

INTRODUCTION: Gooders and Boyer 1986; Kålås et al. 1993; McCaw et al. 1996; Marquiss and Carss 1997. TEXAS DISTRIBUTION: Seyffert 2001; Lockwood and Freeman 2004. POPULATION STATUS: Dugger et al. 1994; Sea Duck Joint Venture 2004a; Davis and Capobianco 2006. DIET: Munro and Clemens 1932; Miller and Barclay 1973; Wood 1987; Kålås et al. 1993; McCaw et al. 1996; Mallory and Metz 1999. RANGE AND HABITATS: Bellrose 1980; Gooders and Boyer 1986; Dugger et al. 1994; Lumsden et al. 1986; Mallory and Metz 1999. REPRODUCTION: Bellrose 1980; Mallory and Lumsden

1994; Coupe and Cooke 1999; Mallory and Metz 1999. APPEARANCE: Palmer 1976b; Bellrose 1980; Mallory and Metz 1999.

RED-BREASTED MERGANSER

Mergus serrator

Red-breasted Mergansers often forage cooperatively. When foraging cooperatively, they form a loose line and herd fish into shallow areas. Groups feeding cooperatively may number up to 100. Snowy Egrets and other wading birds may join Red-breasted Merganser flocks once the fish are pushed into shallows.

TEXAS DISTRIBUTION

Breeding: There is a breeding record for Red-breasted Mergansers (two females with a brood) at Laguna Atascosa NWR in Cameron County. This is phenomenal, considering this species primarily nests in boreal forests and arctic regions. Small numbers may occur along the coast during summer.
Migration and Winter: Migrants occur in East Texas and along the coast. They are rare in West Texas except for the El Paso area, where they are locally common. Wintering birds primarily occur in the Coastal Prairies, especially the Laguna Madre.

Red-breasted Merganser (male). *Photograph by Raymond S. Matlack, February 22, 2007, Bolivar Peninsula, Galveston County, Texas.*

Red-breasted Merganser (female). *Photograph courtesy of Arthur Morris/VIREO, taken January 2001, Bolsa Chica Lagoon, Huntington Beach, Calif.*

TEXAS HARVEST

There is no harvest estimate for Red-breasted Mergansers in Texas. However, harvest is likely low, as mergansers are not prized by hunters.

POPULATION STATUS

Red-Breasted Mergansers likely have a stable population.

DIET

Red-breasted Mergansers forage by diving. They find food by sight or by probing small cavities along the bottom for potential food items. They forage on a wide variety of fish, mollusks, crayfish, and large invertebrates. Gulf toadfish, sheepshead minnows, and snapping shrimp constituted 86 percent of their diet in the Laguna Madre of Texas.

RANGE AND HABITATS

Breeding: Red-breasted Mergansers primarily breed in the boreal forests and tundra zones of North America, Europe, and Asia. Breeding pairs are associated with coastal areas, rivers, and lakes. They nest on the ground. Nests are found in herbaceous vegetation and under brush and low-growing conifer branches. Loose colonies may develop on islands.

Migration and Winter: Red-breasted Mergansers migrate through interior and coastal regions of North America. Most winter in coastal regions of Alaska, Canada, Mexico, and the continental United States. They are also found in the Great Lakes, Great Salt Lake, Salton Sea, and lower stretches of the Colorado River (California). In the Laguna Madre they use sea grass beds, which harbor small fish and shrimp.

REPRODUCTION

Pair Bonds: Red-breasted Mergansers form seasonal pair bonds, beginning in their second winter (or later). Some females are paired by February, and pair bond formation accelerates after March. Males follow their mates to breeding locations. Pair bonds dissolve after females begin incubation.
Nesting: Females line nest bowls with grass litter and down. They lay approximately 1 egg every 1–2 days, and their clutch size is about 10 eggs. Only females incubate. Their incubation period is 30–31 days. They frequently parasitize nests of other Red-breasted Mergansers.
Ducklings: After ducklings leave the nest, females brood them at night and during inclement weather. The amalgamation of several broods with one female is common. Females abandon young before they can fly. Ducklings fledge in less than 65 days.

APPEARANCE

Breeding: Adult males have a ragged crest, and their head and back are black with a green to purple sheen. Their upper neck and rear neck are also black. The forward part of their lower neck is white. Their breast is cinnamon with black spots, and the sides of their breast are black. They have gray sides, which are separated from their iridescent back by a broad, horizontal white line. Their lower back and upper tail are gray, and their belly is white. Adult females have a rusty brown head and upper neck. Their breast is light gray, and their back and sides are gray. Their belly is white.

Adult males have a red bill. Females have a bill that is orange brown with pinkish sides. Both sexes have long, narrow serrated bills. Adult males and females weigh 2.4 and 2.0 pounds, respectively.
Nonbreeding: Both sexes tend to have pale, almost white lower chins but otherwise resemble adult females in breeding plumage.
Immature/subadult: This plumage is acquired in their first winter. Males look like females, but some individuals may have black on their head and dark patches on their body. In females this plumage is similar to their breeding plumage.

SOURCES

INTRODUCTION: Des Lauriers and Brattstrom 1965; Emlen and Ambrose 1979; Titman 1999. TEXAS DISTRIBUTION: Rupert and Brush 1996; White 2002; Lockwood and Freeman 2004. POPULATION STATUS: Titman 1999, Sea Duck Joint Venture 2004b. DIET: Munro and Clemens 1932; Cottam and Uhler 1939; Bowls 1980; Titman 1999. RANGE AND HABITATS: Strong 1912; Weller et al. 1969; Bengston 1970; Stott and Olson 1973; Bellrose 1980; Bowls 1980; Titman 1999; Sea Duck Joint Venture 2004b; Craik and Titman 2009. REPRODUCTION: Weller et al. 1969; Bellrose 1980; Young and Titman 1986; Kahlert et al. 1998; Coupe and Cooke 1999; Titman 1999; Craik and Titman 2009. APPEARANCE: Bellrose 1980, Gooders and Boyer 1986; Titman 1999; Bur et al. 2008.

MASKED DUCK

Nomonyx dominicus

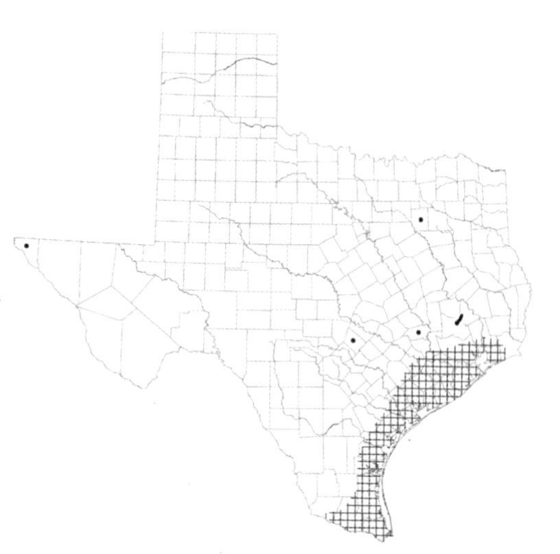

Among native North American waterfowl that breed in the continental United States, Masked Ducks and Muscovy Ducks are the two rarest breeders. Masked Ducks were long suspected of breeding in Texas, but firm evidence was lacking until ducklings were observed in Chambers County in 1967. This was the first US record.

TEXAS DISTRIBUTION

Breeding: Masked Ducks have been documented in the Coastal Prairies and Coastal Sand Plain and in the southeasternmost counties of the South Texas Brush Country. Sightings are irregular and rare. Successful nesting attempts have been documented in Chambers and Live Oak Counties. Breeding is suspected in six more counties, from Hidalgo and Cameron northeast to Jefferson. Nesting in Texas may occur from May to November; this time period is largely inferred from brood sightings. Con-

Masked Duck (male). *Photograph by Mark W. Lockwood, March 31, 1996, Brazos Bend State Park, Fort Bend County, Texas.*

Masked Duck (female). *Photograph by Mark W. Lockwood, March 31, 1996, Brazos Bend State Park, Fort Bend County, Texas.*

sidering the extremely low densities at which they occur, it is unlikely that many Masked Duck nests will ever be found in Texas.

Nonbreeding: Masked Ducks probably do not have a definitive nonbreeding period that holds across their entire range. However, it is likely safe to assume that winter records in Texas are nonbreeders. Many of the larger aggregations (greater than nine) that have been documented in Texas were observed during winter. Only three records are from outside the coastal plain.

Although Masked Ducks have been documented in Texas throughout the year, this does not mean they are a permanent resident. Masked Ducks in the northern portion of their range are reported to wander widely in response to habitat conditions. Thus, there may be periods of time when they are entirely absent from the state. Unfortunately, their habit of occupying ponds with abundant emergent vegetation makes them much harder to detect than most species and clouds their true status in Texas.

There have been four major invasions of Masked Ducks in Texas, or periods during which their numbers were notable. In these periods congregations stay in one location for much longer than do Masked Ducks found in intervening years. The incursions occurred in the 1880s, 1930s, late 1960s/early 1970s, and 1990s. In the early 1990s more than 150 individuals were documented. The reasons for these incursions are unknown but may be related to wetland conditions.

POPULATION STATUS

The population status of Masked Ducks is unknown, but their range may be expanding as a result of climate change.

DIET

Masked Ducks likely forage on seeds (for example, smartweed) and other vegetation. However, most descriptions of their diet are older and relied, at least in part, on gizzard contents. The inclusion of gizzard contents in diet studies biases results toward seeds, which digest considerably slower than invertebrates do.

RANGE AND HABITATS

Breeding: Masked Ducks breed in the Caribbean, along the west-central coast of Mexico (centered around Jalisco and Nayarit), in Central America, and south to Argentina and Uruguay. In the United States they breed in southern Florida and along the Texas coastal plain. They use a wide range of wetland types, including those dominated by floating plants and those with abundant emergent cover (for example, cattails). They also use ephemeral wetlands, shrub-scrub wetlands, rice fields, marshes, forested rivers, mangrove swamps, stock ponds, and open areas of large lakes. They nest overwater, although nesting "near water" is also suggested in some descriptions.
Migration and Winter: Masked Ducks are considered to be nonmigratory. However, their irregular occurrence within the northern portion of their range suggests they are somewhat nomadic. Nomadic movements may explain occasional sightings that occur far outside their regular range. Wetlands used during nonbreeding periods are similar to those used during the breeding season.

REPRODUCTION

Pair Bonds: Masked Ducks are assumed to form short-lived monogamous pair bonds that coincide with the breeding season.
Nesting: Nesting probably coincides with the rainy season in most areas. Very few Masked Duck nests have been found. Only females are believed to construct nest platforms or incubate eggs. Very little down is added to the nest. Females lay four to six eggs, and their incubation period is about 23–24 days. Both renesting and double brooding (two broods in one season) are possible but undocumented.
Ducklings: Females care for ducklings, perhaps leading them back to the nest at night for brooding. The age at which young attain flight is unknown. As with Ruddy Ducks, males may be present on wetlands used for brooding.

APPEARANCE

Breeding: Adult males have a bicolored head; the forward two-thirds, which extends from the bill to well past the eye, is iridescent black, and the remainder is chestnut red. Their overall body color is chestnut red, but it is irregularly marked with black. Their belly is white. Males have a long black tail, which they often hold in an upright position. Adult females have a brownish body with chestnut brown–marked feathers. Their head is pale buff with a black crown, light cheeks, and two parallel facial stripes; one stripe is through the eye and the other is below the eye. Their belly is white. Their tail is long and brownish, and they often hold it in an upright position.

Adult males have a blue bill, and females have a slaty gray bill. Males and females weigh about 13.3 and 12.5 ounces, respectively.

Nonbreeding: In both sexes this plumage is similar to the breeding plumage of adult females.

SOURCES

INTRODUCTION: Johnsgard and Hagemeyer 1969. TEXAS DISTRIBUTION: Johnsgard and Hagemeyer 1969; Blankenship and Anderson 1993; Lockwood 1997; Anderson et al. 1998; Anderson and Tacha 1999; Eitniear 1999, 2010; Benson and Arnold 2003; Lockwood and Freeman 2004; Eubanks et al. 2006. POPULATION STATUS: Rappole et al. 2007. DIET: Bellrose 1980; Johnsgard and Carbonell 1996; Eitniear 1999. RANGE AND HABITATS: Bowman 1995; Howell and Webb 1995; Johnsgard and Carbonell 1996; Anderson and Tacha 1999; Eitniear 1999; Eitniear and Colón-López 2005. REPRODUCTION: Johnsgard and Hagemeyer 1969; Johnsgard 1975; Gomez-Dallmeier and Cringan 1990; Johnsgard and Carbonell 1996; Eitniear 1999. APPEARANCE: Johnsgard 1975; Johnsgard and Carbonell 1996; Eitniear 1999.

RUDDY DUCK

Oxyura jamaicensis

Compared to other waterfowl, Ruddy Ducks have extremely large eggs and a large total clutch weight. A single egg is equivalent to about 12 percent of a female's body weight, and remarkably, the weight of all eggs in a clutch totals about 88 percent of a female's weight. There are even records of individual females having a total clutch weight that exceeded their body weight. For comparison, average egg weight and clutch weight of a Mallard are about 6 percent and 46 percent, respectively, of a female's body weight. The energy required for a Ruddy Duck to produce just one egg exceeds 300 percent of its normal daily energy demands. To meet the energy demands for egg production, female Ruddy Ducks forage voraciously.

TEXAS DISTRIBUTION

Breeding: Ruddy Ducks nest across much of Texas, including the High Plains, Trans-Pecos, South Texas Brush Country, and Coastal Prairies. They are considered casual breeders in the High Plains, and most records are from playa wetlands. In a four-year study conducted in the 1970s, 108 adults and five broods were observed in 12 High

Ruddy Duck (male). *Photograph courtesy of Glenn Bartley/VIREO, taken May 19, 2009, British Columbia, Canada.*

Plains counties. Nesting effort apparently increases when conditions are favorable, particularly in wetlands associated with the Rio Grande.

Migration and Winter: Migrants occur throughout Texas. They arrive in most regions of the state by September or early October. During winter, the vast majority of Ruddy Ducks are found in the Coastal Prairies (TPWD unpublished). Most migrating and wintering Ruddy Ducks depart Texas by mid- to late May.

TEXAS HARVEST

From 1999 to 2006, harvest in Texas averaged 1,610 annually, which was about 6 percent of their annual US harvest.

POPULATION STATUS

From 1994 to 2003 their population was estimated to average 1.1 million annually. However, this estimate is speculative, as much of their breeding range is not surveyed. Ruddy Ducks are believed to have an increasing population.

DIET

Ruddy Ducks primarily forage by diving and searching for prey along the bottom. In California, 80 percent of the diet of breeding adults was midge larvae and leeches. Vegetation such as spikerush and seeds accounted for 38 percent of their winter diet in South Carolina. In spring, midge immatures and brine fly larvae accounted for 95 percent of their diet.

RANGE AND HABITATS

Breeding: The extensive breeding range of Ruddy Ducks covers an area from southern Nunavut south to the Mexican Highlands. They also breed in the Caribbean and in South America, although many authorities give these Ruddy Ducks a subspecies status. They have been introduced to Europe. In North America they reach their peak breeding density in the Prairie Pothole and Parkland Regions. Breeding Ruddy Ducks settle in on semipermanent and permanent freshwater wetlands. Wetlands used for nesting typically have extensive emergent vegetation. Overwater nests are well concealed and typically located in emergent vegetation such as prairie whitetop and cattail. They occasionally nest on the ground in the immediate vicinity of wetlands.

Migration and Winter: Migrants potentially occur on wetlands throughout North America. During winter, Ruddy Ducks are found across much of the United States, Mexico, and Central America. Ruddy Ducks use reservoirs, lakes, aquaculture wetlands, saline lakes, freshwater river deltas, coastal marshes, mangrove swamps, estuaries, and bays. In oxbows located in San Patricio County, they used open water areas 45–84 inches deep.

REPRODUCTION

Pair Bonds: Ruddy Ducks typically form weak, short-lived pair bonds after they arrive on their breeding grounds. Loose, temporary male-female associations that last only a few days have also been described, and females may even switch mates during the breeding season. Less commonly, males have been reported to form simultane-

Ruddy Duck (female) *Photograph by Mark W. Lockwood, April 2, 2009, Balmorhea Lake, Reeves County, Texas.*

ous pair bonds with multiple females, and multiple males may associate with single females. Pair bonds that last through the laying period dissolve in early incubation.
Nesting: Females construct nest platforms. They add very little, if any, down. Only one egg is laid per day. Clutch size varies from seven to eight eggs. Only females incubate. During incubation breaks, they seldom leave the wetland where their nest is located. Their incubation period is about 24 days. Renesting is uncommon. Females frequently lay their eggs in the nests of other ducks.
Ducklings: Females lead ducklings from the nest within 24 hours after they hatch; however, they seldom lead ducklings away from their initial wetland. Females abandon young at about 20 days. Ducklings fledge at 53–67 days.

APPEARANCE

Breeding: Adult males have a black crown and white cheeks. Plumage on the remainder of their upper body is reddish. They have slight, erectable hornlike tufts of feathers on each side of their head and a thick neck, which they widen with inflatable air sacs. Their belly is gray and their tail is black. Adult females have drab, grayish-brown bodies. They have a dark brown crown, light cheeks, and a facial stripe below the eye. Both sexes often sit on the water with their tail in an upright position. The flight of Ruddy Ducks has been described as bumblebee-like because of their very small wings, chunky body, and rapid wing-beat.

Adult males have a broad, bright blue bill. Females have a slaty gray bill. During winter, adult males and females in California both weighed about 1.1 pounds.
Nonbreeding: In both sexes, this plumage is similar to the breeding plumage of adult females except that adult males have whitish cheeks.

SOURCES

INTRODUCTION: Lack 1968; Tome 1984; Woodin and Swanson 1989; Alisauskas and Ankney 1994; Rohwer 1988; Pelayo 2001. TEXAS DISTRIBUTION: Traweek 1978; Pulich 1988; Seyffert 2001; White 2002; Benson and Arnold 2003; Lockwood and Freeman 2004; Baar et al. 2008. TEXAS HARVEST: Kruse 2007. POPULATION STATUS: NAWMP 2004; McNair et al. 2006. DIET: Siegfried 1973; Bellrose 1980; Hoppe et al. 1986; Tome and Wrubleski 1988; Woodin and Swanson 1989; Hohman et al. 1992. RANGE AND HABITATS: Bennett 1938; Low 1941; Miller and Collins 1954; McKnight 1974; White and James 1978; Weller 1988; Johnsgard and Carbonell 1996; Maxson and Riggs 1996; Anderson et al. 2000b; Brua 2001; Baar et al. 2008. REPRODUCTION: Low 1941; Hochbaum 1944; Joyner 1977; Siegfried 1976, 1977; Tome 1991; Solberg and Higgins 1993; Alisauskas and Ankney 1994; Johnsgard and Carbonell 1996; Brua and Machin 2000; Brua 2001; Pelayo 2001. APPEARANCE: Bellrose 1980; Gray 1980; Tome 1984; Johnsgard and Carbonell 1996; Havera 1999a; Brua 2001.

Scientific Names of Animals and Plants Occurring in the Text

COMMON NAME	SCIENTIFIC NAME
Animals	
American Coot	*Fulica americana*
Blue mussel	*Mytilus edulis*
Brook trout	*Salvelinus fontinalis*
Dwarf surf clams	*Mulinia lateralis*
Franklin's Gull	*Larus pipixcan*
Gizzard shad	*Dorosoma cepedianum*
Gulf toadfish	*Opsanus beta*
Herring	*Clupea* spp.
Northern Flicker	*Colaptes auratus*
Osprey	*Pandion haliaetus*
Pileated Woodpecker	*Dryocopus pileatus*
Sheepshead minnow	*Cyprinodon variegates*
Snapping shrimp	*Alpheus* spp.
Snowy Egret	*Egretta thula*
Threadfin shad	*Dorosoma petenense*
Water fleas	*Daphnia* spp., others
Zebra mussels	*Dreissena polymorpha*
Plants	
Alfalfa	*Medicago sativa*
Alkali-bulrush	*Scirpus maritimus* or *Bolboschoenus maritimus*
Alpine chickweed	*Cerastium alpinum*
Arctic bell heather	*Cassiope tetragona*
Arrow grass	*Triglochin* spp.
Arrowhead	*Sagittaria* spp.
Aspen	*Populus tremuloides*
Banana waterlily	*Nymphoides aquatica*
Barnyard grass	*Echinochloa crus-galli*
Beaksedge	*Rhynchospora* spp.

COMMON NAME	SCIENTIFIC NAME
Bearded sprangletop	*Leptochloa fascicularis*
Bermuda grass	*Cynodon* spp.
Bulrush	*Scirpus* spp., *Schoenoplectus* spp., others
Cattail	*Typha* spp.
Chufa	*Cyperus esculentus*
Clover	*Trifolium* spp.
Coontail	*Ceratophyllum* spp.
Corn	*Zea mays*
Cranberry	*Vaccinium* spp.
Delta duck potato	*Sagittaria platyphylla*
Dwarf birch	*Betula glandulosa*
Eelgrass	*Zostera* spp.
Flatsedge	*Cyperus* spp.
Horsetail	*Equisetum* spp.
Hydrilla	*Hydrilla* spp.
Marshhay cordgrass	*Spartina patens*
Mountain cranberry	*Vaccinium vitis-idaea*
Muskgrass	*Chara* spp.
Oats	*Avena sativa*
Olney bulrush	*Scirpus americanus*
Pendant grass	*Arctophila fulva*
Pondweed	*Potamogeton* spp.
Prairie whitetop	*Scolochloa festucacea*
Prickly pear	*Opuntia* spp.
Rice	*Oryza sativa*
Sago pondweed	*Potamogeton pectinatus*
Saltgrass	*Distichlis* spp.
Saltmarsh bulrush	*Scirpus robustus*
Sea lettuce	*Ulva* spp.
Shoalgrass	*Halodule wrightii*
Signalgrass	*Urochloa* spp.
Sorghum (grain sorghum)	*Sorghum* spp.
Soybeans	*Glycine max*
Spruce	*Picea* spp.
Smartweed	*Polygonum* spp.
Spikerush	*Eleocharis* spp.
Water hyacinth	*Eichhornia crassipes*
Western wheatgrass	*Pascopyrum smithii*
Wheat (also winter wheat)	*Triticum* spp.
Widgeongrass	*Ruppia maritima*
Wildcelery	*Vallisneria americana*
Wild rice	*Zizania aquatica*

References

Abraham, K. F., R. L. Jefferies, and C. D. MacInnes. 1996. Why are there so many white geese in North America? *Proc. Int. Waterfowl Symp.* 7:79–92.

Adair, S. E., J. L. Moore, and W. H. Kiel Jr. 1996. Wintering diving duck use of coastal ponds: An analysis of alternative hypotheses. *J. Wildl. Manage.* 60:83–93.

Afton, A. D. 1978. Incubation rhythms and egg temperature of an American Green-winged Teal and a renesting pintail. *Prairie Nat.* 10:115–19.

———. 1980. Factors affecting incubation rhythms of Northern Shovelers. *Condor* 82:132–37.

———. 1984. Influence of age and time on reproductive performance of female Lesser Scaup. *Auk* 101:255–65.

———. 1993. Post-hatch brood amalgamation in Lesser Scaup: Female behavior and return rates, and duckling survival. *Prairie Nat.* 25:227–35.

Afton, A. D., and R. H. Hier. 1991. Diets of Lesser Scaup breeding in Manitoba. *J. Field Ornithol.* 62:325–34.

Afton, A. D., R. H. Hier, and S. L. Paulus. 1991. Lesser Scaup diets during migration and winter in the Mississippi Flyway. *Can. J. Zool.* 69:328–33.

Afton, A. D., and R. D. Sayler. 1982. Social courtship and pairbonding of Common Goldeneyes, *Bucephala clangula,* wintering in Minnesota. *Can. Field-Nat.* 96:295–300.

Aldrich, J. W. 1949. *Migration of some North American waterfowl.* Spec. Sci. Rep., Wildl. 1. Washington, D.C.: USFWS.

Aldrich, J. W., and K. P. Baer. 1970. Status and speciation in the Mexican Duck (*Anas diazi*). *Wilson Bull.* 82:63–73.

Alisauskas, R. T. 1998. Winter range expansion and relationships between landscape and morphometrics of midcontinent Lesser Snow Geese. *Auk* 115:851–62.

Alisauskas, R. T., and C. D. Ankney. 1992. Spring habitat use and diets of midcontinent adult Snow Geese. *J. Wildl. Manage.* 56:43–54.

———. 1994. Costs and rates of egg formation in Ruddy Ducks. *Condor* 96:11–18.

Alisauskas, R. T., C. D. Ankney, and E. E. Klaas. 1988. Winter diets and nutrition of midcontinental Lesser Snow Geese. *J. Wildl. Manage.* 52:403–14.

Alison, R. M. 1975. Breeding biology and behavior of the Oldsquaw (*Clangula hyemalis* L.). *Ornithol. Monogr.* 18:1–52.

Allen, C. E. 1980. Feeding habits of ducks in a green-tree reservoir in eastern Texas. *J. Wildl. Manage.* 44:232–36.

Allen, R. B., P. O. Corr, and J. A. Dorso. 1990. Nesting success and efficiency of waterfowl using nest boxes in central Maine: A management perspective. In *The*

1988 North American Wood Duck Symposium, 291–96. Puxico, Mo.: Gaylord Memorial Laboratory.

Anderson, E. M., J. R. Lovvorn, and M. T. Wilson. 2008. Reevaluating marine diets of Surf and White-winged Scoters: Interspecific differences and the importance of soft-bodied prey. *Condor* 100:285–95.

Anderson, J. T., and D. A. Haukos. 2003. Breeding ground affiliation and movements of Greater White-fronted Geese staging in northwestern Texas. *Southwest. Nat.* 48:365–72.

Anderson, J. T., G. T. Muehl, and T. C. Tacha. 1998. Distribution and abundance of waterbirds in coastal Texas. *Bird Popul.* 4:1–15.

Anderson, J. T., G. T. Muehl, T. C. Tacha, and D. S. Lobpries. 2000a. Wetland use by non-breeding ducks in coastal Texas, U.S.A. *Wildfowl* 51:191–214.

Anderson, J. T., and L. M. Smith. 1999. Carrying capacity and diel use of managed playa wetlands by nonbreeding waterbirds. *Wildl. Soc. Bull.* 27:281–91.

Anderson, J. T., L. M. Smith, and D. A. Haukos. 2000b. Food selection and feather molt by nonbreeding American Green-winged Teal in Texas playas. *J. Wildl. Manage.* 64:222–30.

Anderson, J. T., and T. C. Tacha. 1999. Habitat use by Masked Ducks along the Gulf Coast of Texas. *Wilson Bull.* 111:119–21.

Anderson, M. G., R. B. Emery, and T. W. Arnold. 1997. Reproductive success and female survival affect local population density of Canvasbacks. *J. Wildl. Manage.* 61:1174–91.

Ankney, C. D. 1982. Annual cycle of body weight in Lesser Snow Geese. *Wildl. Soc. Bull.* 10:60–64.

Ankney, C. D., and A. D. Afton. 1988. Bioenergetics of breeding Northern Shovelers: Diet, nutrient reserves, clutch size, and incubation. *Condor* 90:459–72.

Ankney, C. D., and C. D. MacInnes. 1978. Nutrient reserves and reproductive performance of female Lesser Snow Geese. *Auk* 95:459–71.

Anteau, M. J., and A. D. Afton. 2008. Diets of Lesser Scaup during spring migration throughout the upper-midwest are consistent with the spring condition hypothesis. *Waterbirds* 31:97–106.

———. 2009. Lipid reserves of Lesser Scaup (*Aythya affinis*) migrating across a large landscape are consistent with the "Spring Condition" hypothesis. *Auk* 126:873–83.

Armbruster, J. S. 1982. Wood Duck displays and pairing chronology. *Auk* 99:116–22.

Arnold, T. W., M. G. Anderson, R. B. Emery, M. D. Sorenson, and C. N. De Sobrino. 1995. The effects of late-incubation body mass on reproductive success and survival of Canvasbacks and Redheads. *Condor* 97:953–62.

Arnold, T. W., and R. G. Clark. 1996. Survival and philopatry of female dabbling ducks in south-central Saskatchewan. *J. Wildl. Manage.* 60:560–68.

Arnold, T. W., M. D. Sorenson, and J. J. Rotella. 1993. Relative success of overwater and upland Mallard nests in southwestern Manitoba. *J. Wildl. Manage.* 57:578–81.

Austin, J. E., M. J. Anteau, J. S. Barclay, G. S. Boomer, F. C. Rohwer, and S. M. Slattery. 2006. Declining scaup populations: Reassessment of the issues, hypotheses, and research directions. Consensus Rep. Second Scaup Workshop. Bismarck, N.Dak.: USGS, Northern Prairie Wildlife Research Center.

Austin, J. E., C. M. Custer, and A. D. Afton. 1998. Lesser Scaup, *Aythya affinis*. *Birds N. Amer.* 338:1–32.

Austin, J. E., and M. R. Miller. 1995. Northern Pintail, *Anas acuta*. *Birds N. Amer.* 163:1–32.

Austin, J. E., and J. R. Serie. 1991. Habitat use and movements of Canvasback broods in southwestern Manitoba. *Prairie Nat.* 23:223–28.

Austin, J. E., J. R. Serie, and J. H. Noyes. 1990. Diet of Canvasbacks during breeding. *Prairie Nat.* 22:171–76.

Baar, L., R. S. Matlack, W. P. Johnson, and R. B. Barron. 2008. Migration chronology of waterfowl in the Southern High Plains of Texas. *Waterbirds* 31:394–401.

Badzinski, S. S. 2005. Social influences of Tundra Swan activities during migration. *Waterbirds* 28:316–25.

Badzinski, S. S., R. J. Cannings, T. E. Armenta, J. Komaromi, and P. J. A. Davidson. 2008. Monitoring coastal bird populations in BC: The first five years of the Coastal Waterbird Survey (1999–2004). *Brit. Col. Birds* 17:1–35.

Badzinski, S. S., and S. A. Petrie. 2006. Diets of Lesser and Greater Scaup during autumn and spring migration on the lower Great Lakes. *Wildl. Soc. Bull.* 34:664–74.

Baldassarre, G. A., and E. G. Bolen. 1984. Field-feeding ecology of waterfowl wintering on the Southern High Plains of Texas. *J. Wildl. Manage.* 48:63–71.

———. 1986. Body weight and aspects of pairing chronology of Green-winged Teal and Northern Pintails wintering on the Southern High Plains of Texas. *Southwest. Nat.* 31:361–66.

Baldwin, J. R., and J. R. Lovvorn. 1992. Population, diet, food availability, and food requirements of dabbling ducks in Boundary Bay. In *Abundance, distribution and conservation of birds in the vicinity of Boundary Bay, British Columbia*, ed. R. W. Butler, 42–69. Tech. Rep. Ser., No. 155. Ottawa, Ont.: Can. Wildl. Serv.

———. 1994. Expansion of sea grass habitat by the exotic *Zostera japonica*, and its use by dabbling ducks and Brant in Boundary Bay, British Columbia. *Mar. Ecol. Prog. Ser.* 103:119–27.

Ball, I. J., R. L. Eng, and S. K. Ball. 1995. Population density and productivity of ducks on large grassland tracts in northcentral Montana. *Wildl. Soc. Bull.* 23:767–73.

Ballard, B. M., M. T. Merendino, R. H. Terry, and T. C. Tacha. 2001. Estimating abundance of breeding Mottled Ducks in Texas. *Wildl. Soc. Bull.* 29:1186–92.

Ballard, B. M., and T. C. Tacha. 1995. Habitat use by geese wintering in southern Texas. *Southwest. Nat.* 40:68–75.

Ballard, B. M., J. E. Thompson, M. J. Petrie, M. Chekett, and D. G. Hewitt. 2004. Diet and nutrition of Northern Pintail wintering along the southern coast of Texas. *J. Wildl. Manage.* 68:371–82.

Banks, R. C., C. Cicero, J. L. Dunn, A. W. Kratter, P. C. Rasmussen, J. V. Remsen Jr., J. D. Rising, and D. F Stotz. 2004. Forty-fifth supplement to the American Ornithologists' Union Check-list of North American Birds. *Auk* 121:985–95.

Barzen, J. B., and J. R. Serie. 1990. Nutrient reserve dynamics of breeding Canvasbacks. *Auk* 107:75–85.

Batt, B. D. J., ed. 1997. *Arctic ecosystems in peril: Report of the Arctic Goose Habitat*

Working Group. A special publication of the Arctic Goose Joint Venture of the NAWMP. Washington, D.C.: USFWS, and Ottawa, Ont.: Can. Wildl. Serv.

Batt, B. D. J., A. D. Afton, C. D. Ankney, D. H. Johnson, J. A. Kadlec, and G. L. Krapu, eds. 1992. *Ecology and management of breeding waterfowl*. Minneapolis: Univ. Minn. Press.

Bellrose, F. C. 1980. *Ducks, geese and swans of North America*, 3rd ed. Harrisburg, Pa.: Stackpole Books.

Bellrose, F. C., and R. C. Crompton. 1981. Migration speeds of three waterfowl species. *Wilson Bull.* 93:121–24.

Bellrose, F. C., and D. J. Holm. 1994. *The ecology and management of the Wood Duck*. Harrisburg, Pa.: Stackpole Books.

Bellrose, F. C., K. L. Johnson, and T. U. Meyers. 1964. Relative value of natural cavities and nest houses for Wood Ducks. *J. Wildl. Manage.* 28:661–76.

Bengston, S.-A. 1970. Location of nest sites of ducks in Lake M'yvatn area, northeast Iceland. *Oikos* 21:218–29.

Bennett, L. J. 1938. *The Blue-winged Teal: Its ecology and management*. Ames, Iowa: Collegiate Press.

Benson, K. L. P., and K. A. Arnold, eds. 2003. The Texas breeding bird atlas. Tex. A&M Univ. Syst., College Station and Corpus Christi. http://txtbba.tamu.edu.

Bergan, J. F., and L. M. Smith. 1986. Food robbery of wintering Ring-necked Ducks by American Coots. *Wilson Bull.* 98:306–8.

———. 1989. Differential habitat use by diving ducks wintering in South Carolina. *J. Wildl. Manage.* 53:1117–26.

Bergstrom, B. J. 1999. First reported breeding of the Black-Bellied Whistling-Duck in Northern Florida. *Fla. Field Nat.* 27:177–79.

Billard, R. S., and P. S. Humphrey. 1972. Molts and plumages in the Greater Scaup. *J. Wildl. Manage.* 36:765–74.

Bjarvall, A. 1968. The hatching and next-exodus behaviour of Mallard. *Wildfowl* 19:70–80.

Blankenship, T. L., and J. T. Anderson. 1993. A large concentration of Masked Ducks (*Oxyura dominica*) on the Welder Wildlife Refuge, San Patricio County, Texas. *Bull. Tex. Ornithol. Soc.* 26:19–21.

Blohm, R. J. 1979. The breeding ecology of the Gadwall in southern Manitoba. Ph.D. diss., Univ. Wis., Madison.

Bluhm, C. K. 1985. Mate preferences and mating patterns of Canvasbacks (*Aythya valisineria*). *Ornithol. Monogr.* 37:45–56.

Bogiatto, R. J., and J. D. Karnegis. 2006. The use of eastern Sacramento Valley vernal pools by ducks. *Calif. Fish Game* 92:125–41.

Bolen E. G. 1964. Weights and linear measurements of Black-bellied Whistling Ducks. *Tex. J. Sci.* 16:257–60.

———. 1967. The ecology of the black-bellied tree duck in southern Texas. Ph.D. diss., Utah State Univ., Logan.

———. 1971a. Pair bond tenure in black-bellied tree duck. *J. Wildl. Manage.* 35:385–88.

———. 1971b. Some views on exotic waterfowl. *Wilson Bull.* 83:430–34.

Bolen, E. G., and B. W. Cain. 1968. Mixed Wood Duck–tree duck clutch in Texas. *Condor* 70:389–90.

Bolen, E. G., and B. R. Chapman. 1981. Estimating winter sex ratios for Buffleheads. *Southwest. Nat.* 26:49–52.

Bolen, E. G., and B. J. Forsyth. 1967. Foods of the Black-bellied Tree Duck in south Texas. *Wilson Bull.* 79:43–49.

Bolen, E. G., and R. E. McCamant. 1977. Mortality rates for Black-bellied Whistling Ducks. *Bird-Banding* 48:350–53.

Bolen, E. G., B. McDaniel, and C. Cottam. 1964. Natural history of the Black-bellied Whistling Duck (*Dendrocygna autumnalis*) in southern Texas. *Southwest. Nat.* 9:78–88.

Bolen, E. G., and M. K. Rylander. 1983. *Whistling ducks: Zoogeography, ecology, and anatomy.* Tex. Tech Mus. Ser., No. 20. Lubbock: Tex. Tech Univ. Press.

Bolen, E. G., and E. N. Smith. 1979. Notes on the incubation behavior of Black-bellied Whistling Ducks. *Prairie Nat.* 11:119–23.

Bordage, D., and J-P. L. Savard. 1995. Black Scoter, *Melanitta nigra. Birds N. Amer.* 177:1–20.

Botero, J. E., and D. H. Rusch. 1994. Foods of Blue-winged Teal in two neotropical wetlands. *J. Wildl. Manage.* 58:561–65.

Bowles, W. F., Jr., 1980. Winter ecology of Red-breasted Mergansers on the Laguna Madre of Texas. M.S. thesis, Corpus Christi State Univ., Corpus Christi, Tex.

Bowman, M. C. 1995. Sightings of Masked Duck ducklings in Florida. *Fla. Field-Nat.* 23:35.

Boyd, H. 1974. *Waterfowl studies.* Rep. Ser., No. 29. Ottawa, Ont.: Can. Wildl. Serv.

Brakhage, G. K. 1965. Biology and behavior of tub-nesting Canada Geese. *J. Wildl. Manage.* 29:751–71.

Brown, P. W., and M. A. Brown. 1981. Nesting biology of the White-winged Scoter. *J. Wildl. Manage.* 45:38–45.

Brown, P. W., and L. H. Fredrickson. 1997. White-winged Scoter, *Melanitta fusca. Birds N. Amer.* 274:1–28.

Brua, R. B. 2001. Ruddy Duck, *Oxyura jamaicensis. Birds N. Amer.* 696:1–32.

Brua, R. B., and K. L. Machin. 2000. Determining and testing the accuracy of incubation stage of Ruddy Duck eggs by flotation. *Wildfowl* 51:181–89.

Brush, T. 2005. *Nesting birds of a tropical frontier: The Lower Rio Grande Valley of Texas.* College Station: Tex. A&M Univ. Press.

Brush, T., and J. C. Eitniear. 2002. Status and recent nesting of Muscovy Duck (*Cairina moschata*) in the Rio Grande Valley, Texas. *Bull. Tex. Ornith. Soc.* 33:12–14.

Buckley, P. A., S. S. Mitra, and E. S. Brinkley. 2004. Multiple occurrences of Dark-bellied Brant (*Branta* [*bernicla*] *bernicla*) in North America. *N. Amer. Birds* 58:180–85.

Budeau, D., J. T. Ratti, and C. R. Ely. 1991. Energy dynamics, foraging ecology, and behavior of prenesting Greater White-fronted Geese. *J. Wildl. Manage.* 55:556–63.

Buller, R. J. 1955. Ross's Goose in Texas. *Auk* 72:298–99.

Bur, M. T, M. A. Stapanian, G. Bernhardt, and M. W. Turner. 2008. Fall diets of Red-breasted Merganser (*Mergus serrator*) and walleye (*Sander vitreus*) in Sandusky Bay and adjacent waters of western Lake Erie. *Am. Midl. Nat.* 159:147–61.

Burgess, H., and A. Burgess. 1997. Trumpeter Swans once wintered in Texas—why not now? *N. Am. Swans. Bull. Trumpeter Swan Soc.* 26:50–53.

Cain, B. W. 1968. Growth and plumage development of the Black-bellied Tree Duck, *Dendrocygna autumnalis* (Linneaus). M.S. thesis, Tex. A&I Univ., Kingsville.

Cantin, M., J. Bédard, and H. Milne. 1974. The food and feeding of Common Eiders in the St. Lawrence estuary. *Can. J. Zool.* 52:319–34.

Carrière, S., R. G. Bromley, and G. Gauthier. 1999. Comparative spring habitat and food use by two arctic nesting geese. *Wilson Bull.* 111:166–80.

Carroll, J. J. 1932. A change in the distribution of the fulvous tree duck in Texas. *Auk* 49:343–44.

Cely, J. E. 1979. The ecology and distribution of banana water lily and its utilization by Canvasback ducks. *Proc. Annu. Conf. Southeast. Assoc. Fish Wildl. Agencies* 33:43–47.

Chabreck, R. H. 1979. Winter habitat of dabbling ducks—physical, chemical, and biological aspects. In *Waterfowl and wetlands—An integrated review*, ed. T. A. Bookout, 133–42. Madison: Univ. Wis.

Clapp, R. B, M. K. Klimkiewicz, and J. H. Kennard. 1982. Longevity records of North American birds: Gaviidae through Alcidae. *J. Field Ornithol.* 53:81–208.

Clark, R. G., L. G. Sugden, R. K. Brace, and D. J. Nieman. 1988. The relationship between nesting chronology and vulnerability to hunting of dabbling ducks. *Wildfowl* 39:137–44.

Conner, R. N., C. E. Shackelford, D. Saenz, and R. R. Schaefer. 2001. Interactions between nesting Pileated Woodpeckers and Wood Ducks. *Wilson Bull.* 113:250–53.

Cooch, E. G., R. L. Jefferies, R. F. Rockwell, and F. Cooke. 1993. Environmental change and the cost of philopatry: An example in the Lesser Snow Goose. *Oecologia* 93:128–38.

Cooch, E. G., D. B. Lank, A. Dzubin, R. F. Rockwell, and F. Cooke. 1991a. Body size variation in Lesser Snow Geese: Environmental plasticity in gosling growth rates. *Ecol.* 72:503–12.

Cooch, E. G., D. B. Lank, R. F. Rockwell, and F. Cooke. 1991b. Long-term decline in body size in a Snow Goose population: Evidence of environmental degradation? *J. Anim. Ecol.* 60:483–96.

Cooke, F., and C. M. McNally. 1975. Mate selection and colour preferences in Lesser Snow Geese. *Behaviour* 53:151–70.

Cooke, F., R. F. Rockwell, and D. B. Lank. 1995. *The Snow Geese of La Pérouse Bay: Natural selection in the wild.* Oxford: Oxford Univ. Press.

Cooper, J. A. 1978. The history and breeding biology of the Canada Geese of Marshy Point, Manitoba. *Wildl. Monogr.* 61:1–87.

Corcoran, R. M., J. R. Lovvorn, M. R. Bertram, and M. T. Vivion. 2007. Lesser Scaup nest success and duckling survival on the Yukon Flats, Alaska. *J. Wildl. Manage.* 71:127–34.

Cornish, B. J., and D. L. Dickson. 1997. Common Eiders nesting in the western Canadian Arctic. In *King and Common Eiders of the western Canadian Arctic*, ed. D. L. Dickson, 40–50. Occas. Pap. No. 94. Ottawa, Ont.: Can. Wildl. Serv.

Cottam, C. 1939. *Food habits of North American diving ducks*. Tech. Bull. 643. Washington, D.C.: USDA.

Cottam, C., and W. C. Glazener. 1959. Late nesting of water birds in south Texas. *Trans. N. Amer. Wild. Conf.* 24:382–94.

Cottam, C., J. J. Lynch, and A. L. Nelson. 1944. Food habits and management of American sea brant. *J. Wildl. Manage.* 8:36–56.

Cottam, C., and F. M. Uhler. 1937. *Birds in relation to fishes*. Wildl. Res. Manage. Leaflet BS-83. Washington, D.C.: US Bur. Biol. Surv.

Coupe, M., and F. Cooke. 1999. Factors affecting the pairing chronologies of three species of mergansers in southwest British Columbia. *Waterbirds* 22:452–58.

Cowardin, L. M., G. E. Cummings, and P. B. Reed Jr. 1967. Stump and tree nesting by Mallards and black ducks. *J. Wildl. Manage.* 31:229–35.

Cox, R. R., Jr. and A. D. Afton. 1997. Use of habitats by female Northern Pintails wintering in southwestern Louisiana. *J. Wildl. Manage.* 61:435–43.

———. 2000. Predictable interregional movements by female Northern Pintails during winter. *Waterbirds* 23:258–69.

Craik, S. R., and R. D. Titman. 2009. Nesting ecology of Red-breasted Mergansers in a Common Tern colony in eastern New Brunswick. *Waterbirds* 32:282–92.

Custer, C. M., and T. W. Custer. 1996. Food habits of diving ducks in the Great Lakes after the zebra mussel invasion. *J. Field Ornithol.* 67:86–99.

Davis, S., and P. Capobianco. 2006. The Hooded Merganser: A preliminary look at growth in numbers in the United States as demonstrated in the Christmas Bird Count database. *N. Amer. Birds* 60:27–33.

Dean, L. 2005. The amazing story of Trumpeter Swan H99. USFWS, Region 3, Journal Entry, Feb. 18, 2005. www.fws.gov/arsnew/print/print_report.cfm?arskey=14894.

Delnicki, D. 1973. Renesting, incubation behavior, and compound clutches of the Black-bellied Tree Duck in southern Texas. M.S. thesis, Tex. Tech Univ., Lubbock.

———. 1983. Mate changes by Black-bellied Whistling Ducks. *Auk* 100:728–29.

Delnicki, D., E. G. Bolen, and C. Cottam. 1976. An unusual clutch size of the Black-bellied Whistling Duck. *Wilson Bull.* 88:347–88.

Delnicki, D., and K. J. Reinecke. 1986. Mid-winter food use and body weights of Mallards and Wood Ducks in Mississippi. *J. Wildl. Manage.* 50:43–51.

Des Lauriers, J. R., and B. H. Brattstrom. 1965. Cooperative feeding behavior in Red-breasted Mergansers. *Auk* 62:639.

Devries, J. H., R. W. Brook, D. W. Howerter, and M. G. Anderson. 2008. Effects of spring body condition and age on reproduction in Mallards (*Anas platyrhynchos*). *Auk* 125:618–28.

Dickson, K. M., ed. 2000. *Towards conservation of the diversity of Canada Geese (*Branta canadensis*)*. Occas. Pap. No. 103. Ottawa, Ont.: Can. Wildl. Serv.

Donnelly, W. A., and F. G. Whoriskey. 1991. Background-color acclimation of brook trout for crypsis reduces risk of predation by Hooded Mergansers *Lophodytes cucullatpus*. *N. Am. J. Fish. Manage.* 11:206–11.

Doty, H. A., F. B. Lee, A. D. Kruse, J. W. Matthews, J. R. Foster, and P. M. Arnold.

1984. Wood Duck and Hooded Merganser nesting on Arrowwood NWR, North Dakota. *J. Wildl. Manage.* 48:577–80.
Drewien, R. C., and D. S. Benning. 1997. Status of Tundra Swans and Trumpeter Swans in Mexico. *Wilson Bull.* 109:693–701.
Drewien, R. C., A. L. Terrazas, J. P. Taylor, J. M. Ochoa Barraza, and R. E. Shea. 2003. Status of Lesser Snow Geese and Ross's Geese wintering in the Interior Highlands of Mexico. *Wildl. Soc. Bull.* 31:417–32.
Drilling, N., R. Titman, and F. McKinney. 2002. Mallard, *Anas platyryhnchos. Birds N. Amer.* 658:1–44.
Drobney, R. D. 1980. Reproductive bioenergetics of Wood Ducks. *Auk* 97:480–90.
———. 1982. Body weight and composition changes and adaptations for breeding Wood Ducks. *Condor* 84:300–305.
Drobney, R. D., and L. H. Fredrickson. 1979. Food selection by Wood Ducks in relation to breeding status. *J. Wildl. Manage.* 43:109–20.
Drury, W. H., Jr. 1961. Observations on some breeding water birds on Bylot Island. *Can. Field-Nat.* 75:84–101.
Dubovsky, J. A., and R. M. Kaminski. 1992. Waterfowl and American Coot habitat associations with Mississippi catfish ponds. *Proc. Annu. Conf. Southeast. Assoc. Fish Wildl. Agencies* 46:10–17.
DuBowy, P. J. 1985. Feeding ecology and behavior of postbreeding male Blue-winged Teal and Northern Shovelers. *Can. J. Zool.* 63:1292–97.
———. 1996. Northern Shoveler, *Anas clypeata. Birds N. Amer.* 217:1–24.
Duebbert, H. F., and H. A. Kantrud. 1987. Use of no-till winter wheat by nesting ducks in North Dakota. *J. Soil Water Conserv.* 42:50–53.
Duebbert, H. F., and J. T. Lokemoen. 1976. Duck nesting in fields of undisturbed grass-legume cover. *J. Wildl. Manage.* 40:39–49.
Dugger, B. D., K. M. Dugger, and L. H. Fredrickson. 1994. Hooded Merganser, *Lophodytes cucullatus. Birds N. Amer.* 98:1–24.
Duncan, D. C. 1987a. Nesting of Northern Pintails in Alberta: Laying date, clutch size, and renesting. *Can. J. Zool.* 65:234–46.
———. 1987b. Variation and heritability in egg size of the Northern Pintail. *Can. J. Zool.* 65:992–96.
Durham, R. S., and A. D. Afton. 2003. Nest-site selection and success of Mottled Ducks on agricultural lands in southwest Louisiana. *Wildl. Soc. Bull.* 31:433–42.
———. 2006. Breeding biology of Mottled Ducks on agricultural lands in southwestern Louisiana. *Southwest. Nat.* 5:311–16.
Dzubin, A. 1965. A study of migrating Ross' Geese in western Saskatchewan. *Condor* 67:511–34.
Dzubin, A., and J. B. Gallop. 1972. Aspects of Mallard breeding ecology in Canadian parkland and grassland. In *Population ecology of migratory birds: A symposium*, 113–52. Wildl. Res. Rep. 2. Washington, D.C.: USFWS, Bureau Sport Fish. Wildl.
Dzus, E. H., and R. G. Clark. 1997. Overland travel, food abundance, and wetland use by Mallards: Relationships with offspring survival. *Wilson Bull.* 109:504–15.
Eadie, J. M., K. Cheng, and K. Nichols. 1987. Limitations of tetracycline in tracing multiple maternity. *Auk* 104:330–33.

Eadie, J. M., F. P. Kehoe, and T. D. Nudds. 1988. Pre-hatch and post-hatch brood amalgamation in North American Anatidae: A review of the hypotheses. *Can. J. Zool.* 66:1709–21.

Eadie, J. M., M. L. Mallory, and H. G. Lumsden. 1995. Common Goldeneye, *Bucephala clangula. Birds N. Amer.* 170:1–32.

Eadie, J. M., J-P. L. Savard, and M. L. Mallory. 2000. Barrow's Goldeneye, *Bucephala islandica. Birds N. Amer.* 548:1–32.

Earnst, S. L., and T. C. Rothe. 2004. Habitat selection by Tundra Swans on northern Alaska breeding grounds. *Waterbirds* 27:224–33.

Eberhardt, R. T., and M. Riggs. 1995. Effects of sex and reproductive status on diets of breeding Ring-necked Ducks (*Aythya collaris*) in north-central Minnesota. *Can. J. Zool.* 73:392–99.

Edmonds, S. T., and D. S. Stolley. 2008. Population decline of ground-nesting Black-bellied Whistling-Ducks (*Dendrocygna autmnalis*) on islands in southern Texas. *Southwest. Nat.* 53:185–89.

Eggeman, D. R., D. D. Humburg, D. A. Graber, and L. J. Korschgen. 1989. Temporal changes in fall and winter foods of Canada Geese. *Proc. Annu. Conf. Southeast. Assoc. Fish Wildl. Agencies* 43:372–79.

Eichholz, M. W., and J. S. Sedinger. 2006. Staging, migration, and winter distribution of Canada and Cackling Geese staging in interior Alaska. *J. Wildl. Manage.* 70:1308–15.

Eitniear, J. C. 1999. Masked Duck, *Nomonyx dominica. Birds N. Amer.* 393:1–12.

———. 2010. Noteworthy breeding of Masked Ducks in Live Oak County, Texas. *Bull. Tex. Ornithol. Soc.* 43:87–88.

Eitniear, J. C., A. Aragon-Tapia, and J. T. Baccus. 1998. Unusual nesting of the Muscovy Duck *Cairina moschata* in northeastern Mexico. *Tex. J. Sci.* 50:173–75.

Eitniear, J. C., and S. Colón-López. 2005. Recent observations of Masked Duck (*Nomonyx dominica*) in the Caribbean. *Caribb. J. Sci.* 41:861–64.

Elphick, C. S., and L. W. Oring. 1998. Winter management of Californian rice fields for waterbirds. *J. Applied Ecol.* 35:95–108.

Ely, C. R. 1993. Family stability in Greater White-fronted Geese. *Auk* 110:425–35.

Ely, C. R., and A. X. Dzubin. 1994. Greater White-fronted Goose, *Anser albifrons. Birds N. Amer.* 131:1–32.

Ely, C. R., and D. G. Raveling. 1984. Breeding biology of Pacific white-fronted geese. *J. Wildl. Manage.* 48:823–37.

Emlen, S. T., and H. W. Ambrose III. 1970. Feeding interactions of Snowy Egrets and Red-breasted Mergansers. *Auk* 87:164–65.

Engeling, G. A. 1950. The nesting habits of the Mottled Duck in Wharton, Fort Bend, and Brazoria Counties, Texas, with notes on molting and movements. M.S. thesis, Tex. A&M Univ., College Station.

Erskine, A. J. 1959. A joint clutch of Barrow's Goldeneye and Bufflehead eggs. *Can. Field-Nat.* 73:131.

———. 1960. Further notes on interspecific competition among hole-nesting ducks. *Can. Field-Nat.* 74:161–62.

———. 1972. *Buffleheads.* Monogr. Ser., No. 4. Ottawa, Ont.: Can. Wildl. Serv.

Esler, D., T. D. Bowman, K. A. Trust, B. A. Ballachey, T. A. Dean, S. C. Jewett, C. E. O'Clair. 2002. Harlequin Duck population recovery following the 'Exxon Valdez' oil spill: Progress, process and constraints. *Mar. Ecol. Prog. Ser.* 241:271–86.

Esler, D., and S. A. Iverson. 2010. Female Harlequin Duck winter survival 11 to 14 years after the *Exxon Valdez* oil spill. *J. Wildl. Manage.* 74:471–78.

Esslinger, C. G., and B. C. Wilson. 2001. *Gulf Coast Joint Venture: Chenier Plain Initiative.* NAWMP. Albuquerque, N.M.: NAWMP.

Eubanks, T., R. A Behrstock, and R. J. Weeks. 2006. *Birdlife of Houston, Galveston, and the Upper Texas Coast.* College Station: Tex. A&M Univ. Press.

Euliss, N. H., Jr., and S. W. Harris. 1987. Feeding ecology of Northern Pintails and Green-winged Teal wintering in California. *J. Wildl. Manage.* 51:724–32.

Evans, C. D., A. S. Hawkins, and W. H. Marshall. 1952. Movements of waterfowl broods in Manitoba. Spec. Sci. Rep. Wildl., No. 16. Washington, D.C.: USFWS.

Evans, M. R., D. B. Lank, W. S. Boyd, and F. Cooke. 2002. A comparison of the characteristics and fate of Barrow's Goldeneye and Bufflehead nests in nest boxes and natural cavities. *Condor* 104:610–19.

Evard, J. O. 1999. Mallard and Blue-winged Teal philopatry in northwest Wisconsin. *N. Am. Bird Bander* 24:38–42.

Fast, M., R. G. Clark, R. W. Brook, P. L. F. Fast, J-M. Devink, and S. W. Leach. 2008. Eye colour, aging, and decoy trap bias in Lesser Scaup, *Aythya affinis. Can. Field-Nat.* 122:21–28.

Fedynich, A. M., and R. D. Godfrey Jr. 1989. Gadwall pair recaptured in successive winters on the Southern High Plains of Texas. *J. Field Ornithol.* 60:168–70.

Finger, R. S., B. M. Ballard, M. T. Merendino, J. P. Hurst, D. S. Lobpries, and A. M. Fedynich. 2003. Habitat use, movements, and survival of female Mottled Ducks and ducklings during brood rearing. Unpubl. rep., TPWD, Austin.

Fischer, S. A. 1998. A comparison of duck abundance on Conservation Reserve Program and agricultural land in North Dakota. M.S. thesis, La. State Univ., Baton Rouge.

Fisher, F. M., Jr., S. L. Hall, W. R. Welder, B. C. Robinson, and D. S. Lobpries. 1986. An analysis of spent shot in Upper Texas coastal waterfowl wintering habitat. In *Lead poisoning in wild waterfowl, a workshop*, ed. J. S. Feierabend and A. B Russell, 50–60. Washington, D.C.: Natl. Wildl. Fed.

Fleskes, J. P. 1992. Record of a Redhead, *Aythya americana,* laying eggs in a Northern Harrier, *Circus cyaneus,* nest. *Can. Field-Nat.* 106:263–64.

Fleskes, J. P., D. S. Gilmer, and R. L. Jarvis. 2005. Pintail distribution and selection of marsh types at Mendota Wildlife Area during fall and winter. *Calif. Fish Game* 91:270–85.

Flickinger, E. L. 1975. Incubation by a male fulvous tree duck. *Wilson Bull.* 87:106–107.

Flickinger, E. L., D. S. Lobpries, and H. A. Bateman. 1977. Fulvous Whistling Duck population in Texas and Louisiana. *Wilson Bull.* 89:329–31.

Flint, P. L. 2003. Incubation behaviour of Greater Scaup, *Aythya marila*, on the Yukon-Kuskokwim Delta, Alaska. *Wildfowl* 54:97–105.

Flint, P. L., J. B. Grand, T. F. Fondell, and J. A. Morse. 2006. Population dynamics of

Greater Scaup breeding on the Yukon-Kuskokwim Delta, Alaska. *Wildl. Monogr.* 162:1–22.

Fournier, M. A., and J. E. Hines. 2001. Breeding ecology of sympatric Greater and Lesser Scaup (*Aythya marila* and *Aythya affinis*) in the subarctic Northwest Territories. *Arctic* 54:444–56.

Fowler, A. C., J. M. Eadie, and C. R. Ely. 2004. Relatedness and nesting dispersion within breeding populations of Greater White-fronted Geese. *Condor* 106:600–607.

Fredrickson, L. H., and J. L. Hansen. 1983. Second broods in Wood Ducks. *J. Wildl. Manage.* 47:320–26.

Frentress, C. D. 1987. Wildlife of bottomlands: Species and status. In *Bottomland hardwoods in Texas*, 37–57. PWD-RP-7100–133–3/87. Austin: TPWD.

Gammonley, J. H. 1995. Spring feeding ecology of Cinnamon Teal in Arizona. *Wilson Bull.* 107:64–72.

———. 1996. Cinnamon Teal, *Anas cyanoptera. Birds N. Amer.* 209:1–20.

Gammonley, J. H., and L. H. Fredrickson. 1998. Breeding duck populations and productivity on montane wetlands in Arizona. *Southwest. Nat.* 43:219–27.

Gammonley, J. H., and M. E. Heitmeyer. 1990. Behavior, body composition, and foods of Buffleheads and Lesser Scaups during spring migration through the Klamath Basin, California. *Wilson Bull.* 102:672–83.

Gates, J. M. 1962. Breeding biology of the Gadwall in northern Utah. *Wilson Bull.* 74:43–67.

Gates, R. J., D. F. Caithamer, W. E. Moritz, and T. C. Tacha. 2001. Bioenergetics and nutrition of Mississippi Valley Population Canada Geese during winter and migration. *Wildl. Monogr.* 146:1–65.

Gauthier, G. 1987a. Brood territories in Buffleheads: Determinants and correlates of territory size. *Can. J. Zool.* 65:1402–10.

———. 1987b. Further evidence of long-term pair bonds in ducks of the genus *Bucephala. Auk* 104:521–22.

———. 1987c. The adaptive significance of territorial behaviour in breeding Buffleheads: A test of three hypothesis. *Anim. Behav.* 35:348–60.

———. 1988. Factors affecting nest box use by Buffleheads and other cavity-nesting birds. *Wildl. Soc. Bull.* 16:132–41.

———. 1989. The effect of experience and timing on reproductive performance in Buffleheads. *Auk* 106:568–73.

———. 1990. Philopatry, nest site fidelity, and reproductive performance in Buffleheads. *Auk* 107:126–32.

———. 1993a. Bufflehead, *Bucephala albeola. Birds N. Amer.* 67:1–24.

———. 1993b. Feeding ecology of nesting Greater Snow Geese. *J. Wildl. Manage.* 57:216–23.

Gauthier, G., and J. N. B. Smith. 1987. Territorial behaviour, nest site availability, and breeding density in Buffleheads. *J. Anim. Ecol.* 56:171–84.

Gloutney, M. L., R. T. Alisauskas, A. D. Afton, and S. M. Slattery. 2001. Foraging time and dietary intake by breeding Ross's and Lesser Snow Geese. *Oecologia* 127:78–86.

Gomez-Dallmeier, F., and A. Cringan. 1990. *Biology, conservation, and management of waterfowl in Venezuela*. Caracas, Venezuela: Editorial ex Libris.

Gooders, J., and T. Boyer. 1986. *Ducks of North America and the Northern Hemisphere*. New York: Facts on File.

Goudie, R. I., and C. D. Ankney. 1986. Body size, activity budgets, and diets of sea ducks wintering in Newfoundland. *Ecol.* 67:1475–82.

Goudie, R. I., G. J. Robertson, and A. Reed. 2000. Common Eider, *Somateria mollissima. Birds N. Amer.* 546:1–32.

Grand, J. B. 1992. Breeding chronology of Mottled Ducks in a Texas coastal marsh. *J. Field Ornithol.* 63:195–202.

Gray, B. J. 1980. Reproduction, energetic, and social structure of the Ruddy Duck. Ph.D. diss., Univ. Calif., Davis.

Gray, R. L., and R. F. Schultze. 1977. Nesting on water bank lands in Merced County. *Trans. West. Sec. Wildl. Soc.* 13:97–102.

Greenwood, R. J., A. B Sargeant, D. H. Johnson, L. M. Cowardin, and T. L. Shaffer. 1987. Mallard nest success and recruitment in Prairie Canada. *Trans. N. Amer. Wildl. Nat. Resour. Conf.* 52:298–309.

———. 1995. Factors associated with duck nest success in the Prairie Pothole Region of Canada. *Wildl. Monogr.* 128:1–57.

Grieb, J. R. 1970. The shortgrass prairie Canada Goose population. *Wildl. Monogr.* 22:1–49.

Hagar, C. N. 1945. Harlequin Duck on the Texas coast. *Auk* 62:639–40.

Hansen, H. A., P. E. K. Shepard, J. G. King, and W. A. Troyer. 1971. The Trumpeter Swan in Alaska. *Wildl. Monogr.* 26:1–83.

Hanson, H. C. 1965. *The Giant Canada Goose*. Carbondale: S. Ill. Univ. Press.

———. 2006. *The white-cheeked geese: Taxonomy, ecophysiographic relationships, biogeography, and evolutionary considerations*. Vol. 1. Ill. Nat. His. Surv. Blythe, Calif.: AVVAR Books.

Hanson, H. C., and L. L. Eberhardt. 1971. The Columbia River Canada Goose population 1950–1970. *Wildl. Monogr.* 28:1–61.

Haramis, G. M., and D. Q. Thompson. 1985. Density-production characteristics of box-nesting Wood Ducks in a northern greentree impoundment. *J. Wildl. Manage.* 49:429–36.

Harrigal, D., and J. E. Cely. 2004. Black-bellied Whistling-Ducks nest in South Carolina. *Chat* 68:106–8.

Hartke, K. M., and G. R. Hepp. 2004. Habitat use and preferences of breeding female Wood Ducks. *J. Wildl. Manage.* 68:84–93.

Hartke, K. M., K. H. Kriegel, G. M. Nelson, and M. T. Merendino. 2009. Abundance of wigeongrass during winter and use by herbivorous waterbirds in a Texas coastal marsh. *Wetlands* 29:288–93.

Haukos, D. A., S. Martinez, and J. Heltzel. 2010. Characteristics of ponds used by breeding Mottled Ducks on the Chenier Plain of the Texas Gulf Coast. *J. Fish Wildl. Manage.* 1:93–101.

Haukos, D. A., J. E. Neaville, and J. E. Myers. 2001. *Body condition of waterfowl har-*

vested on the upper Gulf Coast of Texas, 1986–2000. Albuquerque, N.M.: Migratory Bird Office, Region 2, USFWS.

Havera, S. P. 1999a. *Waterfowl of Illinois: Abbreviated field guide*. Manual 7. Champaign: Ill. Nat. Hist. Surv.

———. 1999b. *Waterfowl of Illinois: Status and management*. Publ. 21. Champaign: Ill. Nat. Hist. Surv.

Hawkins, A. S. 1945. Bird life of the Texas Panhandle. *Panhandle-Plains Hist. Rev.* 18:110–50.

Heins, M. N. 1984. Ecology and behavior of Black-bellied Whistling-Duck broods in south Texas. M.S. thesis, Tex. Tech Univ., Lubbock.

Heitmeyer, M. E. 1985. Wintering strategies of female Mallards related to dynamics of lowland hardwood wetlands in the upper Mississippi Delta. Ph.D. diss., Univ. Mo., Columbia.

———. 1988. Body composition of female Mallards in winter in relation to annual life cycle events. *Condor* 90:669–80.

Heitmeyer, M. E., and L. H. Fredrickson. 1981. Do wetland conditions in the Mississippi Delta hardwoods influence mallard recruitment? *Trans. N. Amer. Wildl. Natl. Resour. Conf.* 46:44–57.

Heitmeyer, M. E., and P. A. Vohs Jr. 1984. Characteristics of wetlands used by migrant dabbling ducks in Oklahoma, USA. *Wildfowl* 35:61–70.

Hepp, G. R., and F. C. Bellrose. 1995. Wood Duck, *Aix sponsa*. *Birds N. Amer.* 169:1–24.

Hepp, G. R., and J. D. Hair. 1983. Reproductive behavior and pairing chronology in wintering dabbling ducks. *Wilson Bull.* 95:675–82.

Hepp, G. R., R. T. Hoppe, and R. A. Kennamer. 1987. Population parameters and philopatry of breeding female Wood Ducks. *J. Wildl. Manage.* 51:401–4.

Hepp, G. R., and R. A. Kennamer. 1992. Characteristics and consequences of nest-site fidelity in Wood Ducks. *Auk* 109:812–18.

Herring, G., and J. A. Collazo. 2009. Site characteristics and prey abundance at foraging sites used in Lesser Scaup (*Aythya affinis*) wintering in Florida. *Southwest. Nat.* 8:363–74.

Hesse, B. 1980. Archaeological evidence for Muscovy Duck in Ecuador. *Current Anthropol.* 21:139–40.

Hestbeck, J. B. 1993. Overwinter distribution of Northern Pintail populations in North America. *J. Wildl. Manage.* 57:582–89.

Hier, R. H. 1989. Fall weights of Redheads and Ring-necked Ducks in northern Minnesota. *Prairie Nat.* 21:229–33.

Higgins, K. F. 1977. Duck nesting in intensively farmed areas of North Dakota. *J. Wildl. Manage.* 41:232–42.

Higgins, K. F., L. M. Kirsch, A. T. Klett, and H. W. Miller. 1992. *Waterfowl production on the Woodworth Station in south-central North Dakota, 1965–1981*. Resour. Publ. 180. Washington, D.C.: USFWS.

Hildén, O. 1964. Ecology of duck populations in the island group of Valassaaret, Gulf of Bothnia. *Ann. Zool. Fenn.* 1:153–279.

Hines, J. E. 1977. Nesting and brood ecology of Lesser Scaup at Waterhen Marsh, Saskatchewan. *Can. Field-Nat.* 91:248–55.

Hines, J. E., and G. J. Mitchell. 1983. Gadwall nest-site selection and nesting success. *J. Wildl. Manage.* 47:1063–71.

Hipes, D. L., and G. R. Hepp. 1995. Nutrient-reserve dynamics of breeding male Wood Ducks. *Condor* 97:451–60.

Hobaugh, W. C. 1984. Habitat use by Snow Geese wintering in southeast Texas. *J. Wildl. Manage.* 48:1085–96.

Hobaugh, W. C., and J. G. Teer. 1981. Waterfowl use characteristics of flood-prevention lakes in north-central Texas. *J. Wildl. Manage.* 45:16–26.

Hochbaum, G., and G. Ball. 1978. An aggressive encounter between a pintail with a brood and a Franklin Gull. *Wilson Bull.* 90:455.

Hochbaum, H. A. 1944. *The Canvasback on a prairie marsh*, 2nd ed. Harrisburg, Pa.: Stackpole Co.

Hohman, W. L. 1985. Feeding ecology of breeding Ring-necked Ducks wintering in northwestern Minnesota. *J. Wildl. Manage.* 49:546–57.

———. 1986. Incubation rhythms of Ring-necked Ducks. *Condor* 88:290–96.

———. 1991. Incubation rhythm components for three Cinnamon Teal nesting in California. *Prairie Nat.* 23:229–33.

———. 1993. Body composition of wintering Canvasbacks in Louisiana: Dominance and survival implications. *Condor* 95:377–87.

Hohman, W. L., and C. D. Ankney. 1994. Body size and condition, age, plumage quality, and foods of prenesting male Cinnamon Teal in relation to pair status. *Can. J. Zool.* 72:2172–76.

Hohman, W. L., C. D. Ankney, and D. L. Roster. 1992. Body condition, food habits, and molt status of late-wintering Ruddy Ducks in California. *Southwest. Nat.* 37:268–73.

Hohman, W. L., and R. T. Eberhardt. 1998. Ring-necked Duck, *Aythya collaris. Birds N. Amer.* 329:1–32.

Hohman, W. L., and S. A. Lee. 2001. Fulvous Whistling-Duck, *Dendrocygna bicolor. Birds No. Amer.* 562:1–24.

Hohman, W. L., and D. P. Rave. 1990. Diurnal time-activity budgets of wintering Canvasbacks in Louisiana. *Wilson Bull.* 102:645–54.

Hohman, W. L., T. M. Stark, and J. L. Moore. 1996. Food availability and feeding preferences of breeding Fulvous Whistling-Ducks in Louisiana ricefields. *Wilson Bull.* 108:137–50.

Hohman, W. L., D. W. Woolington, and J. H. Devries. 1990. Food habits of wintering Canvasbacks in Louisiana. *Can. J. Zool.* 68:2605–9.

Holbrook, R. S., F. C. Rohwer, and W. P. Johnson. 2000. Habitat use and productivity of Mottled Ducks on the Atchafalaya River Delta, Louisiana. *Proc. Annu. Conf. Southeast. Assoc. Fish Wildl. Agencies* 54:292–303.

Hoppe, R. T., L. M. Smith, and D. B. Wester. 1986. Foods of wintering diving ducks in South Carolina. *J. Field Ornithol.* 57:126–34.

Howell, S. G. N., and S. Webb. 1995. *A guide to the birds of Mexico and northern Central America.* New York: Oxford Univ. Press.

Hubbard, J. P. 1977. The biological and taxonomic status of the Mexican Duck. Bull. No. 16. Albuquerque: N. Mex. Dept. Game Fish.

Huey, W. S. 1961. Comparison of female Mallard with female New Mexican duck. *Auk* 78:428–31.

Hunt, E. G., and W. Anderson. 1966. Renesting of ducks at mountain meadows, Lassen County, California. *Calif. Fish Game* 52:17–27.

James, J. D., and J. E. Thompson. 2001. Black-bellied Whistling-Duck, *Dendrocygna autumnalis. Birds N. Amer.* 578:1–20.

Jehl, J. R., Jr. 2005. Gadwall biology in a hypersaline environment: Is high productivity offset by postbreeding mortality? *Waterbirds* 28:335–43.

Johnsgard, P. A. 1975. *Waterfowl of North America.* Bloomington: Ind. Univ. Press.

———. 1978. *Ducks, geese, and swans of the world.* Lincoln: Univ. Nebr. Press.

Johnsgard, P. A., and M. Carbonell. 1996. *Ruddy Ducks and other stifftails: Their behavior and biology.* Norman: Univ. Okla. Press.

Johnsgard, P. A., and D. Hagemeyer. 1969. The Masked Duck in the United States. *Auk* 86:691–95.

Johnson, D. H., and J. W. Grier. 1988. Determinants of the breeding distribution of ducks. *Wildl. Monogr.* 100:1–37.

Johnson, K. 1995. Green-winged Teal, *Anas crecca. Birds N. Amer.* 193:1–20.

Johnson, W. P., L. Baar, R. S. Matlack, and R. B. Barron. 2010. Hatching chronology of ducks using playas in the Southern High Plains of Texas. *Am. Midl. Nat.* 163:247–53.

Johnson, W. P., R. S. Holbrook, and F. C. Rohwer. 2002. Nesting chronology, clutch size, and egg size in the Mottled Duck. *Wildfowl* 53:155–66.

Johnson, W. P., and P. R. Garrettson. 2010. Band recovery and harvest data suggest additional American Black Duck records from Texas. *Bull. Tex. Ornithol. Soc.* 43:34–40.

Johnson, W. P., and F. C. Rohwer. 1998. Pairing chronology and agonistic behaviors of wintering Green-winged Teal and Mallards. *Wilson Bull.* 110:311–15.

———. 2000. Foraging behavior of Green-winged Teal and Mallards on tidal mudflats in Louisiana. *Wetlands* 20:184–88.

Johnson, W. P., F. C. Rohwer, and M. Carloss. 1996. Evidence of nest parasitism in Mottled Ducks. *Wilson Bull.* 108:187–89.

Jones, J. J., and R. D. Drobney. 1986. Winter feeding ecology of scaup and Common Goldeneye in Michigan. *J. Wildl. Manage.* 50:446–52.

Jorde, D. G., and R. B. Owen Jr. 1990. Foods of black ducks, *Anas rubripes,* wintering in marine habitats of Maine. *Can. Field-Nat.* 104:300–302.

Joyner, D. E. 1973. Interspecific nest parasitism by ducks and coots in Utah. *Auk* 90:692–93.

———. 1976. Effects of interspecific nest parasitism by Redheads and Ruddy Ducks. *J. Wildl. Manage.* 40:33–38.

———. 1977. Behavior of Ruddy Duck broods in Utah. *Auk* 94:343–49.

Kahlert, J., M. Coupe, and F. Cooke. 1998. Winter segregation and timing of pair formation in Red-breasted Merganser *Mergus serrator. Wildfowl* 49:161–72.

Kålås, J. A., T. G. Heggberget, P. A. Bjørn, and O. Reitan. 1993. Feeding behavior and diet of Goosanders (*Mergus merganser*) in relation to salmonid seaward migration. *Aquat. Living Resour.* 6:31–38.

Kaminski, R. M., and H. H. Prince. 1984. Dabbling duck-habitat associations during spring in Delta Marsh, Manitoba. *J. Wildl. Manage.* 48:37–50.

Kamp, M. B., and J. Loyd. 2001. First breeding records of the Black-bellied Whistling Duck for Oklahoma. *Bull. Okla. Ornithol. Soc.* 34:13–17.

Kantrud, H. A. 1986. Western Stump Lake, a major Canvasback staging area in eastern North Dakota. *Prairie Nat.* 18:247–53.

———. 1993. Duck nest success on Conservation Reserve Program land in the prairie pothole region. *J. Soil Water Conserv.* 48:238–42.

Kantrud, H. A., G. L. Krapu, and G. A. Swanson. 1989. *Prairie basin wetlands of the Dakotas: A community profile.* Biol. Rep. 85 (7.28). Washington, D.C.: USFWS.

Kantrud, H. A., and R. E. Stewart. 1977. Use of natural basin wetlands by breeding waterfowl in North Dakota. *J. Wildl. Manage.* 41:243–53.

Keith, L. B. 1961. A study of waterfowl ecology on small impoundments in southeastern Alberta. *Wildl. Monogr.* 6:1–88.

Kennamer, R. A., and G. R. Hepp. 1987. Frequency and timing of second broods in Wood Ducks. *Wilson Bull.* 99:655–62.

Kerbes, R. H. 1994. *Colonies and numbers of Ross' Geese and Lesser Snow Geese in the Queen Maud Gulf Migratory Bird Sanctuary.* Occas. Pap., No. 81. Ottawa, Ont.: Can. Wildl. Serv.

Kessel, B., D. A. Rocque, and J. S. Barclay. 2002. Greater Scaup, *Aythya marila. Birds N. Amer.* 650:1–32.

Kiel, K. H., Jr. 1970. *A release of hand-reared Mallards in south Texas.* MP-968. College Station: Tex. Agric. Exp. Station, Tex. A&M Univ.

Kinney, S. D. 2004. Estimating the population of Greater and Lesser Scaup during winter in off-shore Louisiana. M.S. thesis, La. State Univ., Baton Rouge.

Knutsen, G. A., and J. C. King. 2004. Bufflehead breeding activity in south-central North Dakota. *Prairie Nat.* 36:187–90.

Koons, D. N., and J. J. Rotella. 2003. Comparative nesting success of sympatric Lesser Scaup and Ring-necked Ducks. *J. Field Ornithol.* 74:222–29.

Koskimies, J., and L. Lahti. 1964. Cold-hardiness of newly hatched young in relation to ecology and distribution in ten species of European ducks. *Auk* 81:281–307.

Kraai, K. J. 2003. Late winter feeding habits, body condition, and feather molt intensity of female Mallards utilizing livestock ponds in northeast Texas. M.S. thesis, Tex. A&M Univ., Commerce.

Kramer, G. W., and N. H. Euliss Jr. 1986. Winter foods of Black-bellied Whistling-Ducks in northwestern Mexico. *J. Wildl. Manage.* 50:413–16.

Krapu, G. L. 1974. Foods of breeding pintails in North Dakota. *J. Wildl. Manage.* 38:408–17.

———. 1981. The role of nutrient reserves on Mallard reproduction. *Auk* 98:29–38.

Krapu, G. L., K. J. Reinecke, D. G. Jorde, and S. G. Simpson. 1995. Spring-staging ecology of midcontinent Greater White-fronted Geese. *J. Wildl. Manage.* 59:736–46.

Kruse, K. L., comp. 2007. *Central Flyway harvest and population survey data book.* Denver, Colo.: USFWS.
Kruse, K. L., D. E. Sharp, and K. E. Gamble, comps. 2007. *Light geese in the Central Flyway June 2007.* Denver, Colo.: USFWS.
Lack, D. L. 1968. *Ecological adaptations for breeding in birds.* London: Methuen.
———. 1974. *Evolution illustrated by waterfowl.* New York: Harper and Row.
Landers, J. L., A. S. Johnson, P. H. Morgan, and W. P. Baldwin. 1976. Duck foods in managed tidal impoundments in South Carolina. *J. Wildl. Manage.* 40:721–28.
Lane, S. J., A. Azuma, and H. Higuchi. 1998. Wildfowl damage to agriculture in Japan. *Agr. Ecosyst. Environ.* 70:69–77.
Lank, D. B., P. Mineau, R. F. Rockwell, and F. Cooke. 1989. Intraspecific nest-parasitism and extra-pair copulation in Lesser Snow Geese. *Anim. Behav.* 37:74–89.
Leonard, J. P., M. G. Anderson, H. H. Prince, and R. B. Emery. 1996. Survival and movements of Canvasback ducklings. *J. Wildl. Manage.* 60:863–74.
Leopold, A. S. 1959. *Wildlife of Mexico: The game birds and mammals.* Berkeley: Univ. Calif. Press.
Leopold, F. 1951. A study of nesting Wood Ducks in Iowa. *Condor* 53:209–20.
LeSchack, C. R., and G. R. Hepp. 1995. Kleptoparasitsm of American Coots by Gadwalls and its relationship to social dominance and food abundance. *Auk* 112:429–35.
LeSchack, C. R., S. K. McKnight, and G. R. Hepp. 1997. Gadwall, *Anas strepera. Birds N. Amer.* 283:1–28.
Leslie, J. C., and R. H. Chabreck. 1984. Winter habitat preference of white-fronted geese in Louisiana. *Trans. N. Amer. Wildl. Nat. Resour. Conf.* 49:519–26.
Limpert, R. J. 1980. Homing success of adult Buffleheads to a Maryland wintering site. *J. Wildl. Manage.* 44:905–8.
Limpert, R. J., H. A. Allen Jr., and W. J. L. Sladen. 1987. Weights and measurements of wintering Tundra Swans. *Wildfowl* 38:108–13.
Limpert, R. J., and S. L. Earnst. 1994. Tundra Swan, *Cygnus columbianus. Birds N. Amer.* 89:1–20.
Lindsey, A. A. 1946. The nesting of the New Mexican duck. *Auk* 63:483–92.
Livizey, B. C. 1981. Duck nesting in retired croplands at Horicon National Wildlife Refuge. *J. Wildl. Manage.* 45:27–37.
Lobpries, D. S. 1987. *Whistling duck investigations.* Performance Rep., March 20, Fed. Aid Project W-106-R-13. Austin: TPWD.
Lockwood, M. W. 1997. A closer look: Masked Duck. *Birding* 29:386–90.
———. 2008. Texas Bird Records Committee report for 2007. *Bull. Texas Ornith. Soc.* 41:37–45.
Lockwood, M. W., and B. Freeman. 2004. *The Texas Ornithological Society handbook of Texas birds.* College Station: Tex. A&M Univ. Press.
Lokemoen, J. T. 1966. Breeding ecology of the Redhead duck in western Montana. *J. Wildl. Manage.* 30:668–81.
———. 1967. Flight speed of the Wood Duck. *Wilson Bull.* 79:238–39.

———. 1973. Waterfowl production on stock-watering ponds in the Northern Plains. *J. Range Manage.* 26:179–84.

———. 1991. Brood parasitism among waterfowl nesting on islands and peninsulas in North Dakota. *Condor* 93:340–45.

Lokemoen, J. T., H. F. Duebbert, and D. E. Sharp. 1990. Homing and reproductive habits of Mallards, Gadwalls, and Blue-winged Teal. *Wildl. Monogr.* 106:1–28.

Lokemoen, J. T., and D. E Sharp. 1981. First documented Cinnamon Teal nesting in North Dakota produced hybrids. *Wilson Bull.* 93:403–5.

Longcore, J. R., D. G. McAuley, G. R. Hepp, and J. M. Rhymer. 2000. American Black Duck, *Anas rubripes. Birds N. Amer.* 481:1–36.

Loos, E. R., and F. C. Rohwer. 2004. Laying stage nest attendance and onset of incubation in prairie nesting ducks. *Auk* 121:587–99.

Lovvorn, J. R. 1990. Courtship and aggression in Canvasbacks: Influence of sex and pair-bonding. *Condor* 92:369–78.

Low, J. B. 1941. Nesting of the Ruddy Duck in Iowa. *Auk* 58:506–17.

———. 1945. Ecology and management of the Redhead, *Nyroca americana,* in Iowa. *Ecol. Monogr.* 15:35–69.

Lumsden, H. G., R. E. Page, and M. Gauthier. 1980. Choice of nest boxes by Common Goldeneyes in Ontario. *Wilson Bull.* 92:497–505.

Lumsden, H. G., J. Robinson, and R. Hartford. 1986. Choice of nest boxes by cavity-nesting ducks. *Wilson Bull.* 98:167–68.

Lutmerding, J. A., and A. S. Love. 2011. Longevity records of North American birds. Ver. 2011.2. Laurel, Md.: Patuxent Wildl. Res. Ctr., Bird Banding Lab. www.pwrc.usgs.gov/bbl/longevity/Longevity_main.cfm.

MacCluskie, M. C., and J. S. Sedinger. 1999. Incubation behavior of Northern Shovelers in the subarctic: A contrast to the prairies. *Condor* 101:417–21.

Mallory, M., and K. Metz. 1999. Common Merganser, *Mergus merganser. Birds N. Amer.* 442:1–28.

Mallory, M. L., and H. G. Lumsden. 1994. Notes on egg laying and incubation in the Common Merganser. *Wilson Bull.* 106:757–59.

Mallory, M. L., and P. J. Weatherhead. 1993. Incubation rhythms and mass loss of Common Goldeneyes. *Condor* 95:849–59.

Manlove, C. A., and G. R. Hepp. 2000. Patterns of nest attendance in female Wood Ducks. *Condor* 102:286–91.

Markum, D. E., and G. A. Baldassarre. 1989. Breeding biology of Muscovy Ducks using nest boxes in Mexico. *Wilson Bull.* 101:621–26.

Marquiss, M., and D. N. Carss. 1997. Fish-eating birds and fisheries. *BTO News.* 210/211:6–7.

Martin, K. H., M. S. Lindberg, J. A. Schmutz, and M. R. Bertram. 2009. Lesser Scaup breeding probability and female survival on the Yukon Flats, Alaska. *J. Wildl. Manage.* 73:914–23.

Matteson, S., S. Craven, and D. Compton. 1995. *The Trumpeter Swan*. Coop. Ext. Publ. G3647. Madison: Univ. Wis.

Maxson, S. J., and R. M. Pace III. 1992. Diurnal time activity budgets and habitat use of Ring-necked Duck ducklings in northcentral Minnesota. *Wilson Bull.* 104:472–84.

Maxson, S. J., and M. R. Riggs. 1996. Habitat use and nest success of overwater nesting ducks in westcentral Minnesota. *J. Wildl. Manage.* 60:108–19.

May, M. E., and J. C. Kroll. 1989. Wood Duck nest site selection in eastern Texas. *Proc. Annu. Conf. Southeast. Assoc. Fish Wildl. Agencies* 43:380–88.

McAuley, D. G., and J. R. Longcore. 1989. Nesting phenology and success of Ring-necked Ducks in east-central Maine. *J. Field Ornithol.* 60:112–19.

McCamant, R. E., and E. G. Bolen. 1979. A 12-year study of nest box utilization by Black-bellied Whistling Ducks. *J. Wildl. Manage.* 43:936–43.

McCaw, J. H., III, P. J. Zwank, and R. L. Steiner. 1996. Abundance, distribution, and behavior of Common Mergansers wintering on reservoirs in southern New Mexico. *J. Field Ornithol.* 67:669–79.

McCracken, K. G., W. P. Johnson, and F. H. Sheldon. 2001. Molecular population genetics, phylogeography, and conservation biology of the Mottled Duck (*Anas fulvigula*). *Conserv. Genetics* 2:87–102.

McKinney, F. 1965. The displays of the American Green-winged Teal. *Wilson Bull.* 77:112–21.

McKnight, D. E. 1974. Dry-land nesting Redheads and Ruddy Ducks. *J. Wildl. Manage.* 38:112–19.

McLandress, M. R. 1983. Temporal changes in habitat selection and nest spacing in a colony of Ross' and Lesser Snow Geese. *Auk* 100:335–43.

McLandress, M. R., and I. McLandress. 1979. Blue-phase Ross' Geese and other blue-phase geese in western North America. *Auk* 96:544–50.

McLaren, P. L., and M. A. McLaren. 1982. Migration and summer distribution of Lesser Snow Geese in interior Keewatin. *Wilson Bull.* 94:494–504.

McMahan, C. A. 1970. Food habits of ducks wintering on Laguna Madre, Texas. *J. Wildl. Manage.* 34:946–49.

McNair, D. B., L. D. Yntema, C. Cramer-Burke, and S. L. Fromer. 2006. Recent breeding records and status review of the Ruddy Duck (*Oxyura jamaicensis*) on St. Croix, U.S. Virgin Islands. *J. Caribb. Ornithol.* 19:91–96.

Meanley, B., and A. G. Meanley. 1958. Nesting habitat of the Black-bellied Whistling Duck in Texas. *Wilson Bull.* 70:94–95.

———. 1959. Observations on the fulvous tree duck in Louisiana. *Wilson Bull.* 71:33–45.

Melinchuk, R., and J. P. Ryder. 1980. The distribution, fall migration routes, and survival of Ross's Geese. *Wildfowl* 31:161–71.

Merendino, M. T., D. S. Lobpries, J. E. Neaville, J. D. Ortego, and W. P. Johnson. 2005. Regional differences and long-term trends in lead exposure in Mottled Ducks. *Wildl. Soc. Bull.* 33:1002–8.

Merrill, J. C. 1878. Notes on the ornithology of southern Texas, being a list of birds observed in the vicinity of Fort Brown, Texas, from February, 1876, to June, 1878. *Proc. U.S. Natl. Mus.* 1:118–73.

Michot, T. C. 2000. Comparison of wintering Redhead populations in four Gulf of

Mexico seagrass beds. In *Limnology and aquatic birds: Monitoring, modeling, and management*, ed. F. A. Comin, J. A. Herrara, and J. Ramirez, 243–60. Mérida, Mexico: Univ. Autónoma de Yucatán.

Michot, T. C., J. B. Low, and D. R. Anderson. 1979. Decline of Redhead duck nesting on Knudson Marsh, Utah. *J. Wildl. Manage.* 43:224–29.

Michot, T. C., M. C. Woodin, and A. J. Nault. 2008. Food habits of Redheads (*Aythya americana*) wintering in seagrass beds of coastal Louisiana and Texas, USA. *Acta Zool. Acad. Sci. H.* 54 (Suppl. 1):239–50.

Mickelson, P. G. 1975. Breeding biology of Cackling Geese and associated species on the Yukon-Kuskokwim Delta, Alaska. *Wildl. Monogr.* 45:1–35.

Migoya, R., and G. A. Baldassarre. 1993. Harvest and food habits of waterfowl wintering in Sinaloa, Mexico. *Southwest. Nat.* 38:168–71.

Miller, A. W., and B. D. Collins. 1954. A nesting study of ducks and coots on Tule Lake and Lower Klamath National Wildlife Refuges. *Calif. Fish Game* 40:17–37.

Miller, D. L., F. E. Smeins, and J. W. Webb. 1996. Mid-Texas coastal marsh change (1939–1991) as influenced by Lesser Snow Geese herbivory. *J. Coastal Res.* 12:462–76.

Miller, F. W. 1954. Ross Goose in Texas. *Condor* 56:312.

Miller, L. C., R. M. Whiting Jr., and M. S. Fountain. 2003. Foraging habits of Mallards and Wood Ducks in a bottomland hardwood forest in Texas. *Proc. Annu. Conf. Southeast. Assoc. Fish Wildl. Agencies* 57:160–71.

Miller, S. W., and J. S. Barclay. 1973. Predation in warm water reservoirs by wintering Common Mergansers. *Proc. Annu. Conf. Southeast. Assoc. Fish Wildl. Agencies* 27:243–52.

Mitchell, C. D. 1994. Trumpeter Swan, *Cygnus buccinators*. *Birds N. Amer.* 105:1–23.

Mlodinow, S. G., P. F. Springer, B. Deuel, L. S. Semo, T. Leukering, T. D. Schonewald, W. Tweit, and J. H. Barry. 2008. Distribution and identification of Cackling Goose (*Branta hutchinsii*) subspecies. *N. Am. Birds* 62:344–60.

Moisan, G., R. I. Smith, and R. K. Martinson. 1967. *The Green-winged Teal: Its distribution, migration, and population dynamics*. Spec. Sci. Rep., Wildl. No. 100. Washington, D.C.: USFWS.

Moore, R. L. 1980. Aspects of the ecology and hunting economics of migratory waterfowl on the Texas High Plains. M.S. thesis, Tex. Tech Univ., Lubbock.

Moorman, A. M., T. E. Moorman, G. A. Baldassarre, and D. R. Richard. 1991. Effects of saline water on growth and survival of Mottled Duck ducklings in Louisiana. *J. Wildl. Manage.* 55:471–76.

Moorman, T. E., and P. N. Gray. 1994. Mottled Duck, *Anas fulvigula*. *Birds N. Amer.* 81:1–20.

Morse, T. E., J. L. Jakabosky, and V. P. McCrow. 1969. Some aspects of the breeding biology of the Hooded Merganser. *J. Wildl. Manage.* 33:596–604.

Morse, T. E., and H. M. Wight. 1969. Dump nesting and its effect on production in Wood Ducks. *J. Wildl. Manage.* 33:284–93.

Moser, T. J., ed. 2001. *The status of Ross's Geese*. Arctic Goose Joint Venture Special Publication. Washington, D.C.: USFWS, and Ottawa, Ont.: Can. Wildl. Serv.

———, comp. 2006. *The 2005 North American Trumpeter Swan survey: A cooperative North American survey*. Denver, Colo.: USFWS.

Moser, T. J., R. D. Lien, K. C. VerCauteren, K. F. Abraham, D. E. Anderson, J. G. Bruggink, J. M. Coluccy, D. A. Graber, J. O. Leafloor, D. R. Luukkonen, and R. E. Trost, eds. 2004. *Proceedings of the 2003 International Canada Goose Symposium*. Madison, Wis.: The Symposium.

Mowbray, T. S. 1999. American Wigeon, *Anas americana. Birds N. Amer.* 401:1–32.

———. 2002. Canvasback, *Aythya valisineria. Birds N. Amer.* 659:1–40.

Mowbray, T. S., F. Cooke, and B. Ganter. 2000. Snow Goose, *Chen caerulescens. Birds N. Amer.* 514:1–40.

Mowbray, T. S., C. R. Ely, J. S. Sedinger, and R. E. Trost. 2002. Canada Goose, *Branta canadensis. Birds N. Amer.* 682:1–44.

Mugica Valdes, L. 1993. The rice agroecosystem, Cuban Fulvous Whistling Ducks, and avian conservation. M.S. thesis, Simon Fraser Univ., Vancouver, BC.

Munro, J. A. 1949. Studies of waterfowl in British Columbia: Green-winged Teal. *Can. J. Res.* 27d(3):149–78.

Munro, J. A., and W. A. Clemens. 1932. Food of the American merganser (*Mergus merganser americana*) in British Columbia: A preliminary paper. *Can. Field-Nat.* 46:166–68.

National Audubon Society. 2010. The Christmas Bird Count historical results: Mottled Duck. http://birds.audubon.org/christmas-bird-count.

North American Waterfowl Management Plan. 2004. *Strategic Guidance: Strengthening the Biological Foundation*. NAWMP, Plan Comm. Gatineau, Que.: Can. Wildl. Serv.; Arlington, Va.: USFWS, and México, D.F.: Sec. Medio Ambiente y Recursos Nat.

Nymeyer, L. A. 1975. The Mexican Duck in southcentral New Mexico: Distribution, abundance, habitat. Unpubl. rep., N. Mex. Game Fish Dept. (FW-17-R), Albuquerque.

Oberholser, H. C. 1974. *The bird life of Texas*. Austin: Univ. Tex. Press.

O'Brien, G. P. 1975. *A study of the Mexican Duck (*Anas diazi*) in southeastern Arizona*. Spec. Rep. No. 5. Phoenix: Ariz. Game Fish Dept.

Ogilvie, M. A., and S. Young. 1998. *Wildfowl of the world* (photographic handbook). London: New Holland Publ. Ltd.

Ohlendorf, H. M., and R. F. Patton. 1971. Nesting record of Mexican Duck (*Anas diazi*) in Texas. *Wilson Bull.* 83:97.

O'Kelley, B. L. 1987. Recruitment of Black-bellied Whistling-Ducks in south Texas with special reference to the use of nest boxes. M.S. thesis, Tex. Tech Univ., Lubbock.

Oring, L. W. 1969. Summer biology of the Gadwall at Delta, Manitoba. *Wilson Bull.* 81:44–53.

Palmer, R. S., ed. 1976a. *Handbook of North American birds*. Vol. 2, Part I. New Haven, Conn.: Yale Univ. Press.

———, ed. 1976b. *Handbook of North American birds*. Vol. 3. New Haven, Conn.: Yale Univ. Press.

Paulus, S. L. 1982. Feeding ecology of Gadwall in Louisiana in winter. *J. Wildl. Manage.* 46:483–89.

———. 1983. Dominance relations, resource use, and pairing chronology of Gadwalls in winter. *Auk* 100:947–52.

Pelayo, J. T. 2001. Correlates and consequences of egg size variation in wild Ruddy Ducks (*Oxyura jamaicensis*). M.S. thesis, Univ. Sask., Saskatoon.

Pérez-Arteaga, A., and K. J. Gaston. 2004. Wildfowl population trends in Mexico, 1961–2000: A basis for conservation planning. *Biol. Conserv.* 115:343–55.

Pérez-Arteaga, A., K. J. Gaston, and M. Kershaw. 2002. Population trends and priority conservation sites for Mexican Duck *Anas diazi. Bird Conserv. Int.* 12:35–52.

Perezgasga, F. V. 1999. Wood Duck *Aix sponsa* breeding in the Nazas River, Durango, Mexico. *Continga* 11:13–14.

Peris, S. J., B. Sanchez, and D. Rodriguez. 1998. Range expansion of the Fulvous Whistling-Duck (*Dendrocygna bicolor*) in Cuba, in relation to rice cultivation. *Caribb. J. Sci.* 34:164–66.

Perry, M. C., and F. M. Uhler. 1988. Food habits and distribution of wintering Canvasbacks, *Aythya valisineria,* on Chesapeake Bay. *Estuaries* 11:57–67.

Peters, M. S., and A. D. Afton. 1993. Diets of ring-necked ducks wintering on Catahoula Lake, Louisiana. *Southwest. Nat.* 38:166–68.

Petersen, M. R. 1990. Nest-site selection by Emperor Geese and cackling Canada Geese. *Wilson Bull.* 102:413–26.

Peterson, B., and G. Gauthier. 1985. Nest site use by cavity-nesting birds of the Cariboo Parkland, British Columbia. *Wilson Bull.* 97:319–31.

Peterson, R. T. 1963. *A field guide to the birds of Texas and adjacent states.* 3rd printing. Boston: Houghton Mifflin.

Peterson, S. R., and R. S. Ellarson. 1977. Food habits of Oldsquaws wintering on Lake Michigan. *Wilson Bull.* 89:81–91.

Pierluissi, S. 2006. Breeding waterbird use of rice fields in southwestern Louisiana. M.S. thesis, La. State Univ., Baton Rouge.

Pietz, P. J., G. L. Krapu, D. A. Brandt, and R. R. Cox Jr. 2003. Factors affecting Gadwall brood and duckling survival in prairie pothole landscapes. *J. Wildl. Manage.* 67:564–75.

Pulich, W. M. 1988. *The birds of North Central Texas.* College Station: Tex. A&M Univ. Press.

Prellwitz, D. M. 1987. Canvasback nesting on man-made islands in Montana. *Prairie Nat.* 19:101–2.

Prevett, J. P., and C. D. MacInnes. 1980. Family and other social groups in Snow Geese. *Wildl. Monogr.* 71:1–46.

Ransom, D., Jr., and C. D. Frentress. 2007. Monitoring Texas Wood Ducks with a cooperative nest-box program. *J. Wildl. Manage.* 71:2743–48.

Rappole, J. H., G. W. Blacklock, and J. Norwine. 2007. Apparent rapid range change in South Texas birds: Response to climate change? In *The changing climate of South Texas, 1900–2100*, ed. J. Norwine and K. John, 133–43. Kingsville: CREST-RESSACA, Tex. A&M Univ.–Kingsville.

Raveling, D. G. 1969. Social classes of Canada Geese in winter. 33:304–18.

———. 1988. Mate retention of Giant Canada Geese. *Can. J. Zool.* 66:2766–68.

Ray, J. D., and K. F. Higgins. 1993. Waterfowl use of and production on nesting structures in South Dakota. *Proc. S. Dak. Acad. Sci.* 72:73–86.

Ray, J. D., and H. W. Miller. 1997. A concentration of small Canada Geese in an urban setting at Lubbock, Texas. *Southwest. Nat.* 42:68–73.

Ray, J. D., B. D. Sullivan, and H. W. Miller. 2003. Breeding ducks and their habitats in the High Plains of Texas. *Southwest. Nat.* 48:241–48.

Reed, A., R. Benoit, M. Julien, and R. Lalumière. 1996. *Goose use of the coastal habitats of northeastern James Bay.* Occas. Pap. No. 92. Ottawa, Ont.: Can. Wildl. Serv.

Reed, A., R. Benoit, R. Lalumière, and M. Julien. 1992. *Duck use of the coastal habitats of northeastern James Bay.* Occas. Pap., No. 90. Ottawa, Ont.: Can. Wildl. Serv.

Reed, A., D. H. Ward, D. V. Derksen, and J. S. Sedinger. 1998. Brant, *Branta bernicla. Birds N. Amer.* 337:1–32.

Reimchen, T. E. 1994. Predators and morphological evolution in threespine sticklebacks. *The evolutionary biology of the threespine stickleback,* ed. M. A. Bell and S. A. Foster, 240–76. Oxford: Oxford Univ. Press.

Rhodes, M. J. 1979. Redheads breeding in the Texas Panhandle. *Southwest. Nat.* 24:691–92.

Rhodes, O. E., Jr., T. L. DeVault, and L. M. Smith. 2006. Seasonal variation in carcass composition of American Wigeon wintering in the Southern High Plains. *J. Field Ornithol.* 77:220–28.

Rhymer, J. M. 1988. The effect of egg size variability on thermoregulation of Mallard (*Anas platyrhynchos*) offspring and its implications for survival. *Oecologia* 75:20–24.

Richkus, K. D. 2002. Northern Pintail nest site selection, nest success, renesting ecology, and survival in the intensively farmed prairies of Southern Saskatchewan: An evaluation of the ecological trap hypothesis. Ph.D. diss., La. State Univ., Baton Rouge.

Richkus, K. D., F. C. Rohwer, and M. J. Chamberlain. 2005. Survival and cause-specific mortality of female Northern Pintails in southern Saskatchewan. *J. Wildl. Manage.* 69:574–81.

Ridlehuber, K. T., B. W. Cain, and N. J. Silvy. 1990. Movements, habitat use, and survival of Wood Duck broods in east-central Texas. *Proc. Annu. Conf. Southeast. Assoc. Fish Wildl. Agencies* 44:284–94.

Rienecker, W. C., and W. Anderson. 1960. A waterfowl nesting study on Tule Lake and Lower Klamath National Wildlife Refuges, 1957. *Calif. Fish Game* 46:481–506.

Rigby, E. A. 2008. Recruitment of Mottled Ducks (*Anas fulvigula*) on the upper Texas Gulf Coast. M.S. thesis, Tex. Tech Univ., Lubbock.

Robert, M., D. Bordage, J-P. L. Savard, G. Fitzgerald, and F. Morneau. 2000. The breeding range of the Barrow's Goldeneye in eastern North America. *Wilson Bull.* 112:1–7.

Robertson, D. G., and R. D. Slack. 1995. Landscape change and its effects on the wintering range of a lesser Snow Goose *Chen caerulescens* population: A review. *Biol. Conserv.* 71:179–85.

Robertson, G. J., and R. I. Goudie. 1999. Harlequin Duck, *Histrionicus histrionicus. Birds N. Amer.* 466:1–32.

Robertson, G. J., and J-P. L. Savard. 2002. Long-tailed Duck, *Clangula hyemalis*. *Birds N. Amer.* 651:1–28.

Rofritz, D. J. 1977. Oligocheata as a winter food source for the Old Squaw. *J. Wildl. Manage.* 41:590–91.

Rohwer, F. C. 1985. The adaptive significance of clutch size in prairie ducks. *Auk* 102:354–61.

———. 1988. Inter- and intraspecific relationships between egg size and clutch size in waterfowl. *Auk* 105:161–76.

Rohwer, F. C., and M. G. Anderson. 1988. Female-biased philopatry, monogamy, and the timing of pair formation in migratory waterfowl. *Current Ornithol.* 5:187–221.

Rohwer, F. C., and S. Freeman. 1989. The distribution of conspecific nest parasitism in birds. *Can. J. Zool.* 67:239–53.

Rohwer, F. C., W. P. Johnson, and E. R. Loos. 2002. Blue-winged Teal, *Anas discors*. *Birds N. Amer.* 625:1–26.

Rollo, J. D., and E. G. Bolen. 1969. Ecological relationships of Blue and Green-winged Teal on the High Plains of Texas in early fall. *Southwest. Nat.* 14:171–88.

Rose, P. M., and D. A Scott. 1997. *Waterfowl population estimates*. 2nd ed. Publ. 44. Wageningen, Neth.: Wetlands Intl.

Ross, R. K., S. A. Petrie, S. S. Badzinski, and A. Mullie. 2005. Autumn diet of Greater Scaup, Lesser Scaup, and Long-tailed Ducks on eastern Lake Ontario prior to zebra mussel invasion. *Wildl. Soc. Bull.* 33:81–91.

Rotella, J. J., and J. T. Ratti. 1992. Mallard brood movements and wetland selection in southwestern Manitoba. *J. Wildl. Manage.* 56:508–15.

Rupert, J. R., and T. Brush. 1996. Red-breasted Mergansers, *Mergus serrator*, nesting in southern Texas. *Southwest Nat.* 41:199–200.

Rusch, D. H., M. D. Samuel, D. D. Humburg, and B. D. Sullivan, eds. 1998. *Biology and management of Canada Geese*. Proc. Intl. Canada Goose Symp., Milwaukee, WI.

Rutherford, W. H., ed. 1965. *Description of Canada Goose populations common to the Central Flyway*. N.p.: Central Flyway Waterfowl Council, Tech. Comm.

Ruwaldt, J. J., L. D. Flacke, and J. M. Gates. 1979. Waterfowl pair use of natural and man-made wetlands in South Dakota. *J. Wildl. Manage.* 43:375–83.

Ryder, J. P. 1967. *The breeding biology of the Ross' Goose in the Perry River region, Northwest Territories*. Can. Wildl. Serv. Rep. Ser. 3. Ottawa, Ont.: Can. Wildl. Serv.

———. 1972. Biology of nesting Ross's Geese. *Ardea* 60:185–215.

Ryder, J. P., and R. T. Alisauskas. 1995. Ross' Goose, *Chen rossii*. *Birds N. Amer.* 102:1–28.

Safine, D. E., and M. S. Lindberg. 2008. Nest habitat selection of White-winged Scoters on Yukon-Flats, Alaska. *Wilson. J. Ornithol.* 120:583–93.

Sauer, J. R., J. E. Hines, and J. Fallon. 2008. The North American Breeding Bird Survey, results and analysis, 1966–2007, v. 5.15.2008. Laurel, Md.: USGS, Patuxent Wildl. Res. Ctr. www.mbr-pwrc.usgs.gov/bbs.

Savard, J-P. L. 1985. Evidence of long-term pair bonds in Barrow's Goldeneye. *Auk* 102:389–91.

———. 1987. Causes and functions of brood amalgamation in Barrow's Goldeneye and Bufflehead. *Can. J. Zool.* 65:1548–53.

———. 1988. Use of nest boxes by Barrow's Goldeneye: Nesting success and effect on the breeding population. *Wildl. Soc. Bull.* 16:125–32.

Savard, J-P. L., D. Bordage, and A. Reed. 1998. Surf Scoter, *Melanitta perspicillata. Birds N. Amer.* 363:1–28.

Savard, J-P. L., and M. Robert. 2007. Use of nest boxes by goldeneyes in eastern North America. *Wilson J. Ornithol.* 119:28–34.

Sayler, R. D., and M. A. Willms. 1997. Brood ecology of Mallards and Gadwalls nesting on islands in large reservoirs. *J. Wildl. Manage.* 61:808–15.

Schamber, J. L., P. L. Flint, J. B. Grand, H. M. Wilson, and J. A. Morse. 2009. Population dynamics of Long-tailed Ducks breeding on the Yukon-Kuskokwim Delta, Alaska. *Arctic* 62:190–200.

Schmidt, J. H., E. J. Taylor, and E. A. Rexstad. 2005. Incubation behaviors and patterns of nest attendance in Common Goldeneyes in interior Alaska. *Condor* 107:167–72.

Schmutz, J. K., R. J. Robertson, and F. Cooke. 1983. Colonial nesting of the Hudson Bay eider duck. *Can. J. Zool.* 61:2424–33.

Schubert, C. A., and F. Cooke. 1993. Egg-laying intervals in the Lesser Snow Goose. *Wilson Bull.* 105:414–26.

Schummer, M. L., S. A. Petrie, and R. C. Bailey. 2008. Dietary overlap of sympatric diving ducks during winter on northeastern Lake Ontario. *Auk* 125:425–33.

Scribner, K. T., S. L. Talbot, J. M. Pearce, B. J. Pierson, K. S. Bollinger, and D. V. Derksen. 2003. Phylogeography of Canada Geese (*Branta canadensis*) in western North America. *Auk* 120:889–907.

Sea Duck Joint Venture. 2004a. *Common Merganser (*Mergus merganser*)*. Sea Duck Information Series, Info. Sheet No. 6. Anchorage, Alaska: USFWS, and Sackville, N.B.: Can. Wildl. Serv.

———. 2004b. *Red-breasted Merganser (*Mergus serrator*)*. Sea Duck Information Series, Info. Sheet No. 11. Anchorage, Alaska: USFWS, and Sackville, N.B.: Can. Wildl. Serv.

Sedinger, J. S. 1997. Waterfowl and wetland ecology in Alaska. In *Freshwaters of Alaska: Ecological synthesis*, Ecol. Stud. 119, ed. A. M. Milner and M. W. Oswood, 155–78. New York: Springer-Verlag.

Sell, D. L. 1979. Fall foods of teal on the Texas High Plains. *Southwest. Nat.* 24:373–75.

Semel, B., and P. W. Sherman. 1986. Dynamics of nest parasitism in Wood Ducks. *Auk* 103:813–16.

Sénéchal, H., G. Gauthier, and J-P. L. Savard. 2008. Nesting ecology of Common Goldeneyes and Hooded Mergansers in a boreal river system. *Wilson J. Ornithol.* 120:732–42.

Serie, J. R., and G. A. Swanson. 1976. Feeding ecology of breeding Gadwalls on saline wetlands. *J. Wildl. Manage* 40:69–81.

Serie, J. R., D. L. Trauger, and D. E. Sharp. 1983. Migration and winter distributions of Canvasbacks staging on the upper Mississippi River. *J. Wildl. Manage.* 47:741–53.

Seyffert, K. D. 2001. *Birds of the Texas Panhandle: Their status, distribution, and history.* College Station: Tex. A&M Univ. Press.

Sheeley, D. G., and L. M. Smith. 1989. Tests of diet and condition bias in hunter-killed Northern Pintails. *J. Wildl. Manage.* 53:765–69.

Siegfried, W. R. 1973. Summer food and feeding of the Ruddy Duck in Manitoba. *Can. J. Zool.* 51:1293–97.

———. 1976. Breeding biology and parasitism in the Ruddy Duck. *Wilson Bull.* 88:566–74.

———. 1977. Notes on behaviour of Ruddy Ducks during the brood period. *Wildfowl* 28:126–28.

Singley, J. A. 1892. Texas birds: List of birds observed in Lee County, Texas. Part II in *Geol. Surv. Tex., 4th Ann. Rep.* Austin: Dept. Agr., Insur., Stat., Hist.

Slattery, S. M. 1994. Neonate reserves, growth and survival of Ross'and Lesser Snow Goose goslings. M.S. thesis, Univ. Sask., Saskatoon.

Slattery, S. M., and R. T. Alisauskas. 2007. Distribution and habitat use of Ross's and Lesser Snow Geese during late brood rearing. *J. Wildl. Manage.* 71:2230–37.

Smart, G. 1965. Development and maturation of primary feathers of Redhead ducklings. *J. Wildl. Manage.* 29:533–36.

Smith, C. M., F. Cooke, G. J. Robertson, R. I. Goudie, and W. S. Boyd. 2000. Long-term pair bonds in Harlequin Ducks. *Condor* 102:201–5.

Smith, L. M., R. L. Pederson, and R. M. Kaminski, eds. 1989. *Habitat management for migrating and wintering waterfowl in North America.* Lubbock: Tex. Tech Univ. Press.

Smith, L. M., L. D. Vangilder, and R. A. Kennamer. 1985. Foods of wintering brant in eastern North America. *J. Field Ornithol.* 56:286–89.

Solberg, K. L., and K. F. Higgens. 1993. Over-water nesting by ducks in northeastern South Dakota. *Prairie Nat.* 25:19–22.

Sorenson, M. D. 1991. The functional significance of parasitic egg laying and typical laying in Redhead ducks: An analysis of individual behavior. *Anim. Behav.* 42:771–96.

———. 1993. Parasitic egg laying in Canvasbacks: Frequency, success, and individual behavior. *Auk* 110:57–69.

———. 1998. Patterns of parasitic egg laying and typical nesting in Redhead and Canvasback ducks. In *Parasitic birds and their hosts: Studies in coevolution*, ed. S. I. Rothstein and S. K. Robinson, 357–75. New York: Oxford Univ. Press.

Soulliere, G. J. 1980. Review of Wood Duck nest-cavity characteristics. In *Proc. 1988 N. Am. Wood Duck Symp.*, 153–62. Puxico, Mo.: Gaylord Memorial Laboratory.

Soutiere, E. C., H. S. Myrick, and E. C. Bolen. 1972. Chronology and behavior of American Wigeon wintering in Texas. *J. Wildl. Manage.* 36:752–58.

Sovada, M. A., R. M. Anthony, and B. D. J. Batt. 2001. Predation on waterfowl in arctic tundra and prairie breeding areas: A review. *Wildl. Soc. Bull.* 29:6–15.

Sowls, L. K. 1955. *Prairie ducks: A study of their behavior, ecology, and management.* Harrisburg, Pa.: Stackpole Co.

Speirs, J. M. 1945. Flight speed of the old-squaw. *Auk* 62:135–36.

Stafford, J. D., L. D. Flake, and P. M. Mammenga. 2001. Evidence for double brooding by a Mallard, *Anas platyrhynchos*, in eastern South Dakota. *Can. Field-Nat.* 115:502–4.

Stafford, J. D., M. M. Horath, A. P. Yetter, C. S. Hine, and S. P. Havera. 2007. Wetland use by Mallards during spring and fall in the Illinois and central Mississippi River valleys. *Waterbirds* 30:394–402.

Stahl, P. W., M. C. Muse, and L. Delgado-Espinoza. 2006. New evidence for pre-Columbian Muscovy Duck *Cairina moschata* from Ecuador. *Ibis* 148:657–63.

Stark, R. S. 1978. Morphological differences between Blue-winged and Cinnamon Teal. Migratory Bird Investigations, Prog. Rep., Job No. 10, W-88-R-23. Fort Collins: Colo. Div. Wildl.

Stewart, P. A. 1971. Wood Duck nesting in chimneys. *Auk* 88:425.

Stewart, R. E., and H. A. Kantrud. 1973. Ecological distribution of breeding waterfowl populations in North Dakota. *J. Wildl. Manage.* 37:39–50.

Stieglitz, W. O. 1966. Utilization of available foods by diving ducks on Apalachee Bay, Florida. *Proc. Annu. Conf. Southeast. Assoc. Fish Wildl. Agencies* 20:42–50.

Stolley, D. S, S. T. Edmonds, and C. U. Meteyer. 2008. Mortality of ducklings of the Black-bellied Whistling-Duck (*Dendrocygna autumnalis*) during their premier swim in a hypersaline lake in south Texas. *Southwest. Nat.* 53:230–35.

Stott, R. S., and D. P. Olson. 1973. Food-habitat relationship of sea ducks on the New Hampshire coastline. *Ecol.* 54:996–1007.

Stotts, V. D., and D. E. Davis. 1960. The black duck in the Chesapeake Bay of Maryland: Breeding behavior and biology. *Chesapeake Sci.* 1:127–54.

Stoudt, J. H. 1982. *Habitat use and productivity of Canvasbacks in southwestern Manitoba, 1961–1972.* Res. Publ. 99. Washington, D.C.: USFWS.

Strong, R. M. 1912. Some observations on the life history of the Red-breasted Merganser, *Mergus serrator,* Linn. *Auk* 29:479–88.

Stutzenbaker, C. D. 1988. The Mottled Duck: Its life history, ecology and management. Austin: TPWD.

Sugden, L. G. 1980. Parasitism of Canvasback nests by Redheads. *J. Field Ornithol.* 51:361–64.

Sugden, L. G., and G. Butler. 1980. Estimating densities of breeding Canvasbacks and Redheads. *J. Wildl. Manage.* 44:814–21.

Sullivan, T. M., R. W. Butler, and W. S. Boyd. 2002. Seasonal distribution of waterbirds in relation to spawning Pacific herrring, *Clupea pallasi,* in the Strait of Georgia, British Columbia. *Can. Field-Nat.* 116:366–70.

Suydam, R. S. 2000. King Eider, *Somateria spectabilis. Birds N. Amer.* 491:1–28.

Swanson, G. A., G. L. Krapu, and J. R. Serie. 1979. Foods of laying female dabbling ducks on the breeding grounds. In *Waterfowl and wetlands: An integrated review,* ed. T. A. Bookhout, 47–57. Madison: Univ. Wis.

Swanson, G. A., and M. I. Meyer. 1977. Impact of fluctuating water level on feeding ecology of breeding Blue-winged Teal. *J. Wildl. Manage.* 41:426–33.

Swanson, G. A., M. I. Meyer, and V. A. Adomaitis. 1985. Foods consumed by breeding Mallards on wetlands of south-central North Dakota. *J. Wildl. Manage.* 49:197–203.

Swanson, G. A., M. I. Meyer, and J. R. Serie. 1974. Feeding ecology of Blue-winged Teals. *J. Wildl. Manage.* 38:396–407.

Swepston, D. A. 1979. The Mexican x Mallard duck in Texas. Unpubl. rep., Fed. Aid Project., W-103-R, TPWD, Alpine.

Talent, L. G., R. L. Jarvis, and G. L. Krapu. 1983. Survival of Mallard broods in south-central North Dakota. *Condor* 85:74–78.

Taylor, T. S. 1978. Spring foods of migrating Blue-winged Teal on seasonally flooded impoundments. *J. Wildl. Manage.* 42:900–903.

Thomas, C. 1982. Wintering ecology of dabbling ducks in central Florida. M.A. thesis, Univ. Mo., Columbia.

Thompson, J. D., and G. A. Baldassarre. 1988. Postbreeding habitat preference of Wood Ducks in northern Alabama. *J. Wildl. Manage.* 52:80–85.

———. 1991. Activity patterns of nearctic dabbling ducks wintering in Yucatan, Mexico. *Auk* 108:934–41.

———. 1992. Dominance relationships of dabbling ducks wintering in Yucatan, Mexico. *Wilson Bull.* 104:529–36.

Thompson, J. E., and C. D. Ankney. 2002. Role of food in territory and egg production of Buffleheads (*Bucephala albeola*) and Barrow's Goldeneyes (*Bucephala islandica*). *Auk* 119:1075–90.

Thompson, M. C. 1961. The flight speed of a Red-breasted Merganser. *Condor* 63:265.

Thorpe, P. 2009. *2008–2009 western Central Flyway light goose productivity report.* Denver, Colo.: USFWS.

Tietje, W. D., and J. G. Teer. 1996. Winter feeding ecology of Northern Shovelers on freshwater and saline wetlands in South Texas. *J. Wildl. Manage.* 60:843–55.

Titman, R. D. 1999. Red-breasted Merganser, *Mergus serrator. Birds N. Amer.* 443:1–24.

Tome, M. W. 1984. Changes in nutrient reserves and organ size of female Ruddy Ducks breeding in Manitoba. *Auk* 101:830–37.

———. 1991. Diurnal activity budget of female Ruddy Ducks breeding in Manitoba. *Wilson Bull.* 103:183–89.

Tome, M. W., and D. A. Wrubleski. 1988. Underwater foraging behavior of Canvasbacks, Lesser Scaups, and Ruddy Ducks. *Condor* 90:168–72.

Trauger, D. L. 1974. Eye colour of female Lesser Scaup in relation to age. *Auk* 91:243–54.

Traweek, M. S., Jr. 1978. *Waterfowl production survey.* Fed. Aid Project Rep. No. W106-R-5. Austin: TPWD.

Tremblay, J-P., G. Gauthier, D. Lepage, and A. Desrochers. 1997. Factors affecting nesting success in Greater Snow Geese: Effects of habitat and association with Snowy Owls. *Wilson Bull.* 109:449–61.

Turnbull, R. E., and G. A. Baldassarre. 1987. Activity budgets of Mallards and American Wigeon wintering in east-central Alabama. *Wilson Bull.* 99:457–64.

Turner, B., R. Tomlinson, R. Leyva, and P. Dominguez. 1994. *Wintering populations of Lesser Snow Geese and Ross' Geese in the northern highlands of México, 1988–1990.* Occas. Pap. No. 84. Ottawa, Ont.: Can. Wildl. Serv.

US Department of Agriculture. 2009. *Final environmental assessment of bird damage management in New Mexico.* Albuquerque, N.M.: Anim. Plant Health Inspection Serv., Wildl. Serv. Prog.

US Fish and Wildlife Service. 1978. "Mexican Duck" comes off endangered list. News release, July 26, USFWS, Washington, D.C.

———. 2005. *Migratory bird hunting activity and harvest during the 2003 and 2004 hunting seasons, preliminary estimates*. Washington, D.C.: USFWS.

———. 2006. *Migratory bird hunting activity and harvest during the 1999 and 2000 hunting seasons, final report*. Washington, D.C.: USFWS.

———. 2007a. *Migratory bird hunting activity and harvest during the 2001 and 2002 hunting seasons, final report*. Washington, D.C.: USFWS.

———. 2007b. *Migratory bird hunting activity and harvest during the 2005 and 2006 hunting seasons: Preliminary estimates*. Washington, D.C.: USFWS.

———. 2008. *Analyses of selected mid-winter waterfowl survey data (1955–2008), Region 2 (Central Flyway Region)*. Albuquerque, N.M.: USFWS.

———. 2011. *Waterfowl population status, 2011*. Washington, D.C.: USFWS.

U.S. Geological Survey. 2010. Bands across North America. USGS, USFWS. www.flyways.us/surveys-and-monitoring/banding-and-marking-programs/bands-across-america.

Vermeer, K. 1968. Ecological aspects of ducks nesting in high densities among larids. *Wilson Bull.* 80:78–83.

———. 1981. Food and populations of Surf Scoters in British Columbia. *Wildfowl* 32:107–16.

———. 1982. Food and distribution of three *Bucephala* species in British Columbia waters. *Wildfowl* 33:22–30.

Vermeer, K., and N. Bourne. 1984. The White-winged Scoter diet in British Columbia waters: Resource partitioning with other scoters. In *Marine birds: Their feeding ecology and commercial fisheries relationships*, ed. D. N. Nettleship, G. A. Sanger, and P. F. Springer, 30–38. Ottawa, Ont.: Can. Wildl. Serv.

Vest, J. L., R. M. Kaminski, A. D. Afton, and F. J. Vilella. 2006. Body mass of Lesser Scaup during fall and winter in the Mississippi Flyway. *J. Wildl. Manage.* 70:1789–95.

Vrtiska, M. P., and S. Sullivan. 2009. Abundance and distribution of Lesser Snow and Ross's Geese in the Rainwater Basin and central Platte River Valley of Nebraska. *Great Plains Res.* 19:147–55.

Wahle, L. C., and J. S. Barclay. 1993. Changes in Greater Scaup foods in Connecticut. *Northeast Wildl.* 50:69–76.

Walker, J., M. S. Lindberg, M. C. MacCluskie, M. J. Petrula, and J. S. Sedinger. 2005. Nest survival of scaup and other ducks in the boreal forest of Alaska. *J. Wildl. Manage.* 69:582–91.

Ward, D. H., A. Reed, J. S. Sedinger, J. M. Black, D. V. Derksen, and P. M. Castelli. 2005. North American Brant: Effects of changes in habitat and climate on population dynamics. *Global Change Biol.* 11:869–80.

Warner, K., D. Nieman, J. Solberg, F. Roetker, R. Walters, S. Durham, and K. Kraai. 2007. Fall inventory of mid-continent white-fronted geese. Unpubl. report, Can. Wildl. Serv., Saskatoon, Sask.

Warren, S. M., A. D. Fox, A. Walsh, and P. O'Sullivan. 1992. Age of first pairing and breeding among Greenland White-fronted Geese. *Condor* 94:791–93.

Wayland, M., and D. K. McNicol. 1994. Movements and survival of Common Goldeneye broods near Sudbury, Ontario, Canada. *Can. J. Zool.* 72:1252–59.

Wege, M. L., and D. G. Raveling. 1984. Flight speed and directional responses to wind by migrating Canada Geese. *Auk* 101:342–48.

Weller, M. W. 1957. Growth, weights, and plumages of the Redhead, *Aythya americana*. *Wilson Bull.* 69:5–38.

———. 1959. Parasitic egg laying in the Redhead (*Aythya americana*) and other North American Anatidae. *Ecol. Monogr.* 29:333–65.

———. 1964. Distribution and migration of the Redhead. *J. Wildl. Manage.* 28:63–102.

———. 1965. Chronology of pair formation in some nearctic *Aythya* (Anatidae). *Auk* 82:227–35.

———, ed. 1988. *Waterfowl in winter*. Minneapolis: Univ. Minn. Press.

Weller, M. W., D. L. Trauger, and G. L. Krapu. 1969. Breeding birds of the West Mirage Islands, Great Slave Lake, N.W.T. *Can. Field-Nat.* 83:344–60.

Wells-Berlin, A. M., H. H. Prince, and T. W. Arnold. 2005. Incubation length of dabbling ducks. *Condor* 107:926–29.

White, D. H., and D. James. 1978. Differential use of freshwater environments by wintering waterfowl of coastal Texas. *Wilson Bull.* 90:99–111.

White, M. 2002. *Birds of northeast Texas*. College Station: Tex. A&M Univ. Press.

Whiting, R. M., Jr., and J. P. Cornes. 2009. Estimating waterfowl densities in a flooded forest: A comparison of methods. *Southeast. Nat.* 8:47–62.

Whyte, R. J., and E. G. Bolen. 1985. Corn consumption by wintering Mallards during morning field-flights. *Prairie Nat.* 17:71–78.

Williams, C. K., M. D. Samuel, V. V. Baranyuk, E. G. Cooch, and D. Kraege. 2008. Winter fidelity and apparent survival of Lesser Snow Goose populations in the Pacific Flyway. *J. Wildl. Manage.* 72:159–67.

Williams, T. O., E. G. Cooch, R. L. Jeffries, and F. Cooke. 1993. Environmental degradation, food limitation, and reproductive output: Juvenile survival in Lesser Snow Geese. *J. Anim. Ecol.* 62:766–77.

Wilson, B. C. 2007. *Gulf Coast Joint Venture: Mottled Duck Conservation Plan*. NAWMP. Albuquerque, N.M.: NAWMP.

Wishart, R. A. 1983. Pairing chronology and mate selection in the American Wigeon (*Anas americana*). *Can. J. Zool.* 61:1733–43.

Wood, C. C. 1987. Predation of juvenile Pacific salmon by the Common Merganser (*Mergus merganser*) on Eastern Vancouver Island. I: Predation during the seaward migration. *Can. J. Fish Aquat. Sci.* 44:941–49.

Woodin, M. C. 1996. Wintering ecology of Redheads (*Aythya americana*) in the western Gulf of Mexico region. *Gibier Faune Sauvage/Game Wildl.* 13:653–65.

Woodin, M. C., and T. C. Michot. 2002. Redhead, *Aythya americana*. *Birds N. Amer.* 695:1–40.

Woodin, M. C., and G. A. Swanson. 1989. Foods and dietary strategies of prairie-nesting Ruddy Ducks and Redheads. *Condor* 91:280–87.

Woodyard, E. R., and E. G. Bolen. 1984. Ecological studies of Muscovy Ducks in Mexico. *Southwest. Nat.* 29:453–61.

Wormington, A., and J. H. Leach. 1992. Concentrations of migrant diving ducks

at Point Pelee National Park, Ontario, in response to invasion of zebra mussels, *Dreissena polymorpha*. *Can. Field-Nat.* 106:376–80.
Yerkes, T. 1998. The influence of female age, body mass, and ambient conditions on Redhead incubation constancy. *Condor* 100:62–68.
———. 2000. Nest-site characteristics and brood-habitat selection of Redheads: An association between wetland characteristics and success. *Wetlands* 20:575–80.
Yerkes, T., and T. Kowalchuk. 1999. Use of artificial nesting structures by Redheads. *Wildl. Soc. Bull.* 27:91–94.
Yocum, C. F., and M. Keller. 1961. Correlation of food habits and abundance of waterfowl, Humbolt Bay, California. *Calif. Fish Game* 47:41–53.
Young, A. D., and R. D. Titman. 1986. Costs and benefits to Red-breasted Mergansers nesting in tern and gull colonies. *Can. J. Zool.* 64:2339–43.
Zicus, M. C. 1984. Pair separation in Canada Geese. *Wilson Bull.* 96:129–30.
———. 1990. Renesting by a Common Goldeneye. *J. Field Ornithol.* 61:245–48.
Zicus, M. C., and S. K. Hennes. 1988. Cavity nesting waterfowl in Minnesota. *Wildfowl* 39:115–23.
Zwank, P. J., P. M. McKenzie, and E. B. Moser. 1988. Fulvous Whistling-Duck abundance and habitat use in southwestern Louisiana. *Wilson Bull.* 100:488–94.
Žydelis, R. D., Esler, W. S. Boyd, D. L. LaCroix, and M. Kirk. 2006. Habitat use by wintering Surf and White-winged Scoters: Effects of environmental attributes and shellfish aquaculture. *J. Wildl. Manage.* 70:154–62.

Index